PANDAS – DEAD END OR DEAD WRONG?

AND ELEVEN OTHER SHORT STORIES

FROM THE FRONTIERS OF BIOSCIENCE: 2014

JOHN R. SPEAKMAN

About the author and the articles

UK born, John R. Speakman is a 1000 talents 'A' Professor at the Institute of Genetics and Developmental Biology, Chinese Academy of Sciences in Beijing, China, and a Professor in the Institute of Biological and Environmental Sciences at the University of Aberdeen, in the UK. He has lived in Beijing, China since 2011. In 2013 he also became a features writer for the Chinese monthly popular science magazine 'Newton'.

Acknowledgements

Thanks to Mary for all her work to get these articles compiled and published as a book. I am also grateful to Lina Zhang who encouraged me to write for Newton, and to Newton magazine for publishing the Chinese versions and allowing me to retain copyright on the English versions of the articles.

Cover photos: John Speakman

Contents

January	Tool use – how low can it go?
February	Fears from the past
March	Cheap formations
April	Flashing too early
May	Lazy? It could be genetic, and there could be a cure around the corner!
June	Crossing the boundary
July	Youthful blood
August	Tibetans have ancient genes for running in thin air
September	Stimulating memories
October	Cheetahs show why walking is more expensive than running
November	Do you need to be smart to produce art
December	Pandas – dead end or dead wrong?

Tool use – how low can it go?

January 2014

What separates humans from animals? Until the 1960s the use and manufacture of tools was thought to be one of the fundamental differences between humans and other animals. That distinction all changed when Jane Goodall, who researched chimp populations in Africa, observed a chimp putting a blade of grass down a hole in a termite mound. It turned out that the chimp was 'fishing' for termites, which grab on to the invading implement with their strong jaws, and are then removed by the chimp to be eaten. Termite fishing has since been observed in many chimp troops. In fact, apart from ourselves, chimps are the most prolific tool users: using stones to crack open nuts, using wood to cut fruits into smaller bite sized pieces, using sticks to defend themselves against predators, making sponges from chewed leaves to collect water and fashioning sharpened sticks to poke into holes to catch sleeping bush babies. Since Goodall's first observations, the number of species that have been observed to use and fabricate 'tools' has steadily expanded. Among the other great apes, Bonobos and Orang utans have been observed using large leaves as umbrellas to protect themselves from both sun and rain. In one of the more amazing observations of tool use, in 2004 a female gorilla was observed using a stick to test the depth of water in a newly created pool, that had been made by elephants. Descriptions of her behaviour show that her use of the stick was very goal directed. She first waded into the pool on 2 legs without the stick, but the water rapidly came up to her waist. She then turned around and left the pool, grabbed a stick and re-entered the pool, using the stick to probe the water in front of her to see how deep it was, or how stable the ground was at the bottom of the pool.

Ok, so it isn't that surprising that our nearest evolutionary neighbours have acquired some tool use skills. In fact, as the list of mammals that use tools has grown in size it has really been among animals that one might have guessed would have some tool use ability. Many other primates use tools. Dolphins have been observed to put sponges onto the ends of their noses to protect themselves from damage when probing in sand, and more remarkably they use empty conch shells to trap fish. Elephants break off twigs and use them to swat flies off areas of their bodies that they can't reach with their trunks alone.

Why do we find it unsurprising that these animals use tools? The answer is that what these animals have in common is their brains. They all have large brains for their body sizes. Among birds the story is the same. Birds have been shown to have some remarkable tool making and using abilities. The most prolific of which is the New Caledonian Crow from the pacific islands of New Caledonia. In one experiment these birds were faced with the problem of recovering a food treat from the bottom of a glass tube that was longer than their beaks. One bird actually bent a wire into a small hook that it could put down the tube to recover the item. In another experiment the birds had to use a short stick, to recover a longer stick from a box. The long stick was long enough to recover a meat treat from another box, but the short stick wasn't. This so-called 'meta-tool' task, of using one tool to get another tool, is a really sophisticated problem and yet Caledonian crows were able to work out what they needed to do in a relatively short time, and execute the actions in the appropriate order. Importantly the birds didn't solve the task by trial and error learning. Four of seven crows presented with this problem solved it on the first attempt! If the short and long sticks were reversed in location the birds never attempted to recover the short stick, they just went straight to the reward box with the long stick and got the meat. This performance was equal to that of chimps and gorillas faced with the same problem. Crows it turns out, and other birds like parrots that have been observed to use tools, have particularly large brains for their size. The association between brain size and tool use seems obvious.

The use of tools requires sophisticated reasoning power and ability to foresee the consequences of a given course of action. Although we cannot access their thoughts, animals using tools seem to have in mind a clear goal for their actions. The chimp really is 'fishing' for termites – not randomly pushing sticks into holes. The crow is sequentially recovering the large stick with an end goal in mind, not just poking around at anything. These behaviours only seem feasible if you have a big enough brain to do the complex reasoning they require.

So, if I was to ask you what animal you think most likely to have joined the tool using club most recently, then the crocodile would probably not be at the top of your candidate list. Crocodiles and alligators are not blessed highly when it comes to the brain department. A large 300 kg alligator has a brain about the size of a walnut, weighing less than 10g. In comparison humans weighing 70 kg have brains weighing around 1300 g. In absolute terms the crocodiles brain is less than 1% the size of that of a human. Of course, size isn't everything, and other factors such as the numbers of neurons, neuronal connections and the extent of surface folding of the cortex have all been previously suggested as more important than size, in terms of characterising the brains capacities and its relationship to intelligence – yet, however you calculate it, crocodiles and alligators (indeed all reptiles) never make it anywhere near the top of the pile. It isn't very surprising that among mammals and birds nearly 100% of them show some form of parental care behaviour but these sophisticated forms of behaviour are observed in less than 5% of reptiles. So the report last month of crocodiles and alligators using tools comes as a really big surprise.

To be clear alligators and crocodiles have not been observed using tools in quite as sophisticated ways that have been observed in chimps and crows. However, they have been observed using sticks as lures to attract wading birds like herons and egrets. These birds nest in trees and build their nests from sticks. So the crocodiles and alligators balance sticks on their noses and then wait for a nest building bird to come along and take the bait. The first observations of this behaviour were made back in 2007 on a crocodile in an Indian zoo, by the first author of the paper. Several similar observations were made at alligator parks in Florida. However, whether this was really pre-mediated tool use, with a goal in mind, is open to some debate. First, crocodiles and alligators routinely lie around motionless in shallow water for hours at a time – looking pretty much like a large stick themselves. Second, sticks also float around in water. Sooner or later a stick and a croc are going to end up together, and sooner or later after that a bird looking for a stick will find one perched on a crocodiles nose….and snap, it will become crocodile dinner. However, this could just be a random chance event and doesn't necessarily mean that the crocodile knew what it was doing – far less had done it with the goal in mind to lure a bird to its demise. In fact the crocodile may not even have been aware that the stick was present.

However, the authors followed up these initial casual observations by more systematically studying alligators at 4 different sites in the USA for a full year, making observations monthly throughout the winter, and weekly throughout the spring and summer. Two of these sites were right next to sites where wading birds had nesting sites, and two were in the same bodies of water but about 2km from the nesting sites. The authors reasoned that if stick use by alligators was more than random it would be more prevalent at the sites where wading birds were known to nest, than at sites further away. This proved to be the case. Stick use was significantly more prevalent at the sites where the birds were nesting. However, that wasn't all. The researchers also observed that the use of sticks was restricted to the period of the year when the birds were building their nests – March to June, and wasn't observed at other times of year. This could of course still be random. Perhaps the birds choose nesting places where sticks are more frequently found and hence at these sites they are more likely by chance to end up on an alligators nose. Plus perhaps sticks broken from trees or falling into the water from old nests are more prevalent in spring. However, the observers discounted these effects because there were not more sticks in the water at the sites near the rookeries. In fact if anything there were less because the birds readily collected them up to build nests. In addition, the maximum numbers of sticks falling from abandoned nests occurred in autumn, after the nesting period was over, and when alligators never used them. This really does seem like the alligators were deliberately using the sticks as lures to capture the birds.

In fact this is not the only sophisticated form of behaviour that crocodilians and alligators have been observed performing. As noted above less than 5% of reptiles have evolved advanced parental care behaviours, but in fact among crocodilian species this number rises to 25%. Perhaps even more disturbing is the suggestion that they may have co-operative hunting strategies, with large aggregations of alligators in shallow water involved in feeding frenzies on fish. Clearly our impressions of these animals as moronic solitary killers needs some reappraisal. They may be more sophisticated then we imagine.

This then raises a wider issue of just how much brain you need to engage in advanced goal directed behaviour like tool use. The walnut sized alligator brain is very small. But actually at 10g it is considerably larger than that of a Caledonian crow which has a brain weighing on average just 7g. Clearly the alligator also has a much bigger body (at 300kg it is about 500x bigger than the crow) and in general bigger animals have bigger brains. But exactly why animals with bigger bodies should 'need' a bigger brain isn't exactly obvious. So maybe there are two routes to becoming clever and evolving sophisticated behaviours like tool use. The first is to develop a big brain for your body size (this is what humans and crows have done). But maybe an alternative route is to just get bigger. This is because as animals get larger their brains get larger and they perhaps have more and more residual capacity in their brains that can be devoted to doing sophisticated intelligent things. Perhaps then it isn't a surprise then that these behaviours are observed in crocodilians that are the largest extant reptiles, with, in absolute terms the largest reptile brains. What then does this imply for the intelligence of their extinct relatives the dinosaurs, some of which were 10 to 100x larger than a large croc?

Fears from the past

February 2014

My friend Zhou's wife is afraid of spiders. Every time she sees one she becomes almost hysterical until he has caught it and removed it from their apartment. Other small arthropods, like cockroaches, she doesn't like, but that is a dislike rather than a fear. It is just the spiders that create a real fear in her. However, there is no obvious rationale for this fear. No spiders found in Beijing are poisonous. In addition poisonous spiders are not found in her hometown. Speaking to her it seems that she never had any traumatic experiences in her childhood relating to spiders. As far as she is aware it is just something she has always had. "I think it's in my DNA" she says when asked about it. "Maybe my grandmother had a bad experience with spiders once and she has passed it down to me as well".

It is an interesting explanation for her phobia, but for most professional biologists this 'grandmother' explanation would be unpalatable. The reason is that such an explanation violates the central dogma of molecular biology. That is genetic information is coded in our DNA. To get the information out, it is transcribed into RNA, which is then translated into a protein by reading the RNA code on a ribosome. The flow of information is strictly in one direction from DNA to RNA to protein. The central dogma states that information cannot flow in the opposite direction. The DNA sequence can be changed – but the changes (mutations) happen at random. So the grandmother explanation for Zhou's wife's spider fear is untenable, because it implies that when her grandmother experienced her supposed traumatic spider events, she was somehow able to manipulate the DNA code specifically so that the same fear would be experienced by her descendents. There is no known mechanism to allow this non-random mutation of the DNA code.

The idea that we can inherit characteristics that are acquired during the lifetime of our antecedents was first proposed in the 1700s and was particularly promoted by the French scientist Jean-Baptiste Lamarck, and as a result it is sometimes called Lamarckism. In fact the idea was extremely popular in the 1800's, and even Darwin subscribed to it. However, as our understanding of the mechanism of inheritance emerged, and the way the DNA-RNA-Protein system works it became clear that inheritance of acquired characteristics is unfeasible. Moreover, there is lots of evidence to suggest such inheritance does not occur. For example, if someone loses a leg in a car accident, and subsequently has children – there is no risk that those children will be born with missing (or even slightly shorter) limbs. The acquired characteristic (a missing leg) cannot be passed back into the DNA code and affect any children they might have. Plus, as the evolutionary biologist Richard Dawkins has pointed out: for over 5000 years some religious groups have been removing the foreskin from male children's penises before they mature and have children themselves. Yet despite such a systematic and widespread mutilation for hundreds of generations, members of this religion are not born today without foreskins. They do not ever acquire the characteristic that has been imposed on all their male relatives for multiple generations into the past.

One argument against these examples is that perhaps the reason the acquired characteristic is not subsequently inherited is that it gives the individual no advantage. Lamarckism implies only the acquisition of things if they are of benefit. If a person loses a leg in a car accident there would be no benefit to their offspring to be born with absent limbs. In the 1920s many experiments were performed on rats to see if they could acquire a trait that was advantageous. One particularly famous set of experiments was performed by Harvard biologist William McDougall. The experiments he did involved putting rats into a tank of water with two possible escape routes. One of these escape routes was lit brightly, and the other was unlit. If the rats chose the exit that was well lit, then it got an electric shock. The measure he used was how many trials it took for the rats to learn to use the darker exit route. There were clear advantages to the rats to learn the task more quickly. McDougall trained rats and then bred them, and then trained their offspring and bred them, and continued this process for 32 generations which took 15 years to complete. He found that the offspring of the rats that were trained on the task subsequently learned the same task faster, and as the generations went on the rats became better and better. This was not a small effect. The actual numbers are remarkable. On average the first 8 generations (generations 1 to 8) took 56 sessions to learn the task, while the last 8 generations (25 to 32) only took 20 sessions. Subsequent attempts to replicate this effect included a 20 year long study by Agar and colleagues which found effectively the same effects. However, a confusing result was that in both experiments (and others) the control groups also improved their learning ability at the same time as the group experiencing training! This is a surprising result that remains poorly explained. Indeed some have used it as evidence of reincarnation – suggesting that the control and trained offspring were equally likely to receive the souls of the original trained rats when they died. Whatever the explanations of the effect it was clear that there was no difference due to actual experience of the training. So Larmarckism and the inheritance of the acquired advantageous characteristic did not appear to be correct – in accord with the 'central dogma' that emerged sometime later, once the role of DNA, RNA and proteins in the process of inheritance had been elucidated.

Given this background, a paper published recently in *Nature Neuroscience* describes some startling discoveries. Two scientists, Brian Dias and Kerry Ressier, from Emory

University in Atlanta, USA, studied mice exposed to the chemical acetophenone. This chemical has a distinctive smell that is a bit like that of cherries. The chemical is detected by a specific odorant receptor. They exposed male mice to this smell, at the same time as giving them small but painful electric shocks. The mice soon came to associate together the smell and the pain. Eventually simply exposing the mice to the smell alone was enough to evoke a response which involved shuddering in fear of the anticipated pain. So far pretty unremarkable. The ability of animals to associate things together like this has been known since Pavlov's dogs salivated at the ringing of bells in the 1890's, the discovery of which won him the Nobel prize in 1904. However, what is remarkable is that the same fear characteristics were also observed in the offspring of the trained mice. These offspring had never ever experienced acetophenone in their entire lives. Yet when it was wafted into their cage they shuddered in fear. With William McDougall and Agar's experiments in mind it is reasonable to ask what were the controls doing? The answer is not much. The authors had two sets of controls. Ones that were not exposed to the chemical – and they showed no reaction. But interestingly mice that had been exposed to a different smell attached to a shock also showed no response to acetophenone, showing that it was a really specific response. This is sort of like being afraid of spiders but not cockroaches. Amazingly a third generation of mice also 'inherited' the response. Plus the fear response was also observed in mice conceived by artificial insemination using sperm donated by the mice conditioned to be afraid of the chemical. This eliminates any possibility that the males were somehow imparting the information to the females they mated with, who then passed on the information to their offspring. The information 'be afraid of acetophenone' was somehow encoded in the sperm – and sperm do not have much else in them than a ball of DNA!

This work has some big implications but also raises enormous questions. In particular: what is the mechanism? How is it possible to break the central dogma in such a dramatic fashion? The researchers found that the mice that were trained to be afraid of the odour had changes in their brains linked to the odour perception system. Notably neurons carrying the odour receptor specific for acetophenone were more abundant when compared with the control mice. These changes were also present in their offspring. It is quite credible to believe that repeated exposure to a given chemical might be linked to changes in the levels of receptor for that chemical in the brain. There is no mystery here. The problem is to understand how it is possible that such an effect could be inherited? How could information flow back to the DNA? The researchers suggest that rather than the DNA sequence itself being affected what actually happens is that the DNA has marks on it that indicate whether a given gene should be transcribed or not. These marks are produced by methylation of the

DNA – adding a methyl group onto either cytosine or adenine bases. The level of methylation in a given gene might then be an indication of how important it is to transcribe it. In fact it is well established that increased rates of cytosine methylation (generally in the promoter region upstream of a gene rather than in the gene itself) reduce gene expression. The point is that information could be imparted to the DNA without actually changing its sequence. But was there any evidence to support this idea? Actually Kessler and Dias found that the sperm of mice sensitised to acetophenone had fewer methylation marks, compared to control mice, on their DNA, specifically in the gene for the acetophenone odour receptor. Perhaps then this reduced level of methylation indicates that this gene needs to be transcribed more during development.

But there are some enormous gaps in this explanation. First, we know nothing about how the fear response to the stimulus manages to change the methylation status of exactly the gene that is linked to that stimulus, and specifically how it manages to do this in the sperm cells. Then, assuming the methylation change is the way that it is being signalled, we also don't know why or how this methylation difference is interpreted as meaning 'be afraid of acetophenone'. Why, for example, is it not interpreted as 'hey there are lots of delicious cherries around, and this is what they smell like, you should be really happy when you smell this'? Indeed it would be interesting if they had used another group of mice and paired the same smell to something pleasurable. Would that response also be inherited in their offspring? Would the same methylation marks have occurred? And perhaps most intriguingly what if they then mated a female mouse from the 'afraid' line, with a male mouse from the 'pleasurable' line? This work definitely raises many more questions than it answers, but importantly it starts to chip away at one of the central tenets of modern biology. We don't know how it works, but this study suggests that the central dogma is not quite so strict as it is generally presumed to be.

And maybe Zhou's wife is right. Maybe her grandmother did have a bad experience with spiders and it is now in her DNA – or more correctly, if it is methylation that causes the effect, ON her DNA. Will their daughter be affected by the same fear? That we also don't know, but the researchers are currently working to find out for how many generations in the mice the fear of acetophenone persists. If this really is something that drives human behaviour then it is clear that the landscape of our phobias and fears may depend on a complex cocktail of past experience stretching into times long before we were born. We really could be a bundle of fears from the past.

Cheap formations

March 2014

The winter is over and thoughts are turning now towards the spring, and the coming summer – not only among us humans, but also among animals as well. In particular migrant bird species will shortly begin preparations for their long migrations around the globe. Among them will be the populations of black cranes and the bar-headed geese that migrate each year to breed at Qinghai lake, the biggest lake in China, which is located in a depression in the Qinghai-Tibetan plateau about 100km north west of Xining. The cranes spend their winters in Yunnan and Guizhou provinces, migrating about 3000 km north in the spring. Many bar-headed geese, however, winter much further south in India, and face a more gruelling journey – not only because it is longer, but because between their wintering grounds at sea level, and the lake where they breed in the summer at around 3000m, stands the enormous Himalayan mountains. This formidable barrier requires that birds climb to at least 5000m to get over it. Climbing to this height by flying poses a problem because the air density at this elevation is only about half that at sea level, making it more energetically costly for birds to generate lift from their wing beats. Plus the partial pressure of oxygen is correspondingly lower as well, potentially impinging on their ability to extract sufficient oxygen to support their flight costs. Until a couple of years ago the dynamics of how the birds achieve this feat were elusive – but a landmark paper in Proceedings of the National Academy of the USA in 2011 showed us just how they do it. The paper involved tracking the birds using GPS loggers and showed that they typically pass over the Himalayas in a single day, climbing continuously from their sea level departure points up to 6000 m in about 6-7 hours. What was perhaps most surprising is that during the day there are winds that blow up the mountain range from the south, and the birds could easily hitch a ride on these winds to minimise their costs. However, the researchers found that instead of doing this they waited until the relative calm of the night, and hence had to sustain the climbs completely under their own power.

One characteristic of the flights of both the geese and the cranes is that they are often seen to fly in formations. The shapes of the formations vary but typically cranes and geese fly slightly behind and to one side of another bird in front of them. This leads to patterns that are often in the shape of the western letter 'V' and so they are sometimes called 'V' formations. Why birds fly in such formations has been a matter of speculation for many decades, and two basic ideas have been proposed to explain the behaviour. The first is that it is connected to orientation. The journey by the bar-headed geese over the Himalayas, for example, involves finding and crossing through several high altitude passes.

Using such passes can reduce by several thousand meters the altitude that the geese need to climb to. Bar headed geese may live up to 20-30 years and so some geese will have made the trip north from India many times before. Hence the V may be led by more experienced geese that have made the journey several times previously. These geese may lead less experienced geese through the best routes, saving everyone energy by avoiding the need to climb to higher altitudes (although occasionally this seems to happen as small groups have been heard flying over the high mountain tops including Mount Everest). This idea however raises several difficult questions. Why for example is it necessary to fly in a V shape? Why do they not fly directly behind each other, which would fulfil the orientation function equally well? Plus it seems that flying in these types of formation is a characteristic only of larger birds like geese and cranes. Yet many smaller birds migrate, often in groups, but they do so in a sort of chaotic pattern rather than in the regimented shape that characterises flights of geese and cranes. Presumably these small birds also have demands in terms of orientation and experience – but they appear to solve this problem without the need to resort to flying in V shapes.

The second idea then is that this behaviour has something to do with saving energy and hence reducing the costs of the migration flights. You will have seen similar behaviour among the speed skaters in the recent winter Olympics and the cyclists during the summer Olympics. In the team event the skaters (and cyclists) go along in a tight formation with one person in the lead and three others tucked in a tight line directly behind. A major part of the cost of skating, and cycling, is overcoming the drag from the air that the skater is travelling through. By positioning themselves closely behind another person the skaters in positions 2,3 and 4 all benefit because this behaviour reduces the drag they experience. The drag reduction makes them more efficient, but it is tiring for the leader who still has to pay the full cost of the drag due to the air resistance. Consequently, in skating and cycling teams they rotate the person at the front to equalise the benefits. Close observations of migrating birds flying in V formations suggest that they also use this sort of team work with the lead bird rotating with others every now and then. However, if the function of flying in formation is to reduce drag then it is reasonable to ask why the birds don't fly in straight lines, with one bird immediately behind the other, since this pattern seems to minimise the amount of drag experienced by the followers. We can be pretty sure that if a team of Olympic skaters or cyclists adopted a V formation, rather than a tight line, they would not be a strong contender for a medal. The reason is that there is a big difference between flying and terrestrial modes of locomotion like skating and cycling. The biggest difference is that while drag due to air resistance is also a major part of the cost of flying, there is also a major cost involved in generating a lift force to keep the birds airborne. Cyclists and skaters don't have to pay this cost because they are supported by the reaction force from the ground they are standing on. Perhaps flying slightly offset from the bird in front doesn't minimise drag, but instead allows the following bird to somehow reduce its lift costs.

In fact there seem to be good reasons to think that this is indeed the case. Flying birds shed from their wings a vortex of circulating air. If you were to look at a bird from behind and could see the circulating air, then the vortex shed from the right wing rotates anticlockwise, while the vortex shed from the left wing rotates clockwise. So right in the middle behind the birds back, in both the trailing vortexes, the air is moving in a downward direction. But out to either side the air is circulating upwards. It becomes obvious now why birds don't fly immediately behind each other because if they did they would always be in the air that rotating in a downwards direction – a downwash. This would provide an additional down force that they would need to overcome, in addition to sustaining their own weight, so it would make their flight more rather than less costly. But by positioning themselves to one side they can take advantage of the airstream that is rotating upwards (an upwash) and gain some free lift, making their flight cheaper. This actually works. People have flown light aircraft slightly offset from each other and have shown that the trailing aircraft uses much less fuel than the leading one if it flys in the upwash from the vortex shed off the wing of the leading plane. The savings can be up to 18% and the effect is regularly used by military aircraft to minimise fuel costs. This isn't feasible for passenger flights, however, because the close proximity that planes would need to fly in to get the benefits raises safety concerns. A problem is that to get the benefit from the vortex you need to fly in a very exact location, and this is made more difficult in a bird because the wing is beating up and down so the vortex is also oscillating about in space and time. It has been suggested that this is the reason why flying in formation is only observed in large birds. Wing beat frequency in flying animals is inversely related to body size. Insects may beat their wings at 50-60 times a second (Hertz). Little birds like sparrows beat their wings at about 12 to 15 Hertz, but geese beat their wings much more slowly at 4-5 Hertz and cranes beat their wings even more slowly at 3 to 4 Hertz. Tracking the position of the vortex may therefore be impossible in small birds, explaining why they do not normally fly in any type of formation.

Although this seems a compelling idea over the years many investigators have suggested that the problem of tracking the trailing vortex, to position themselves optimally is too difficult even for large birds and hence the energy saving idea cannot be correct – the so-called 'dynamic wake problem'. However, the idea the V formations act to save birds energy has been confirmed by an amazing study of Bald ibis which was recently published in Nature. The study was performed by Dr Steve Portugal and colleagues and involved a flock of tame ibises that were raised at a zoo in Vienna in Austria. The birds were imprinted onto humans after they hatched, and at 4 months of age were trained to fly behind a powered parachute (a paraplane). Since the birds would follow the paraplane the researchers were able to get a birds eye view of what was going on literally from inside the formation. The paper includes some cool video images as supplementary material taken from inside the paraplane where you can imagine what it is like to be a bird leading a V formation.

The hard data was collected by instrumenting the birds with small GPS loggers and then taking them on a long 'migration' flight from Austria to Italy. The GPS loggers reported the birds positions at 5 Hz frequency resulting in a positional accuracy of 0.3 m and a velocity accuracy of 0.1 m/s. This allowed the positions of the birds relative to the optimal positioning to be continuously monitored and a heat map of their position generated. Further logging equipment allowed the wing beat patterns to be identified and hence the phase relationships of the different birds to be elucidated. The authors could then match together the spatial and wingbeat data to track exactly what each bird was doing in relation to the bird in front. With accelerometry data being generated from each of 14 birds at 300 Hertz and positional data generated at 5 Hertz interpolated up to 300 Hertz the data flow was on the high side – hence the whole paper was based on just a 7 minute segment of formation flying. These data convincingly showed that when flying in a 'V' formation the birds were remarkably good at remaining in a relatively small target area downstream from the bird in front, and that this area corresponded closely to the optimal position predicted by fixed wing aerodynamic theory. Perhaps most surprisingly, however, the birds were able to maintain their wing beats in temporal phase with the bird in front to maximise the upwash that they were flying in. This involved rather complex modulation of the phasing of the wing beat relative to the bird in front, according to exactly where they were spatially. Indeed if for some reason the birds ended up directly behind the one in front they started to flap their wings completely out of phase with the bird in front to minimise the effect of the downwash region that is immediately behind a flying bird. How they achieve this feat however remains unclear. Portugal and colleagues suggest that birds flying in V formations are acutely aware of the wake structures around them and are able to continuously adjust their positions and the phasing of their wing beats relative to their nearest neighbour to maximise their energy savings. However, if the savings in flight are anything near to what is achieved in fuel economy of aircraft flying in formation, the birds may not need to consciously track the wakes, but simply feel their way by where their demands are minimised. After all if I cycle immediately behind someone and take advantage of the drag effect I don't need to know anything about the dynamics of airflow around the body in front of me – only that it is easier to ride there than if we ride side by side. Perhaps flying in formation flight is the same, some areas just feel naturally nice to be in.

Flashing too early

April 2014

When I first came to China, several years back, I made a concentrated effort to learn the Chinese language. I bought myself lots of books, I invested in a set of 'flash cards' that have the characters on them, I listened to lots of DVDs, and in the end I paid out for a personal tutor to train me. Part way through my struggle to learn Chinese my daughter found a really interesting set of DVDs that she suggested I should try out. I already had loads of DVDs and so I told her that I didn't need another set, but she insisted that these were different – and indeed, when I checked out the product, they were different. These were DVDs that you listen to when you are asleep! I didn't really understand how this might work, because when I am awake I hear lots of Chinese being spoken, and it doesn't magically enter into my ears and then later come out of my mouth as a string of coherent language. So hearing Chinese when asleep seemed unlikely to have the same effect.

Nevertheless, in our time constrained world, the idea that we can use those 'wasted hours' of the night time to actually do something productive is a powerful attractor, so I investigated a little more closely. In fact the recordings they play you in the night don't include any Chinese language at all. They supposedly work by a process of self hypnosis. The idea is that our subconscious mind has an image of ourselves. The suggestion is that our difficulty learning things comes because our own self image prevents us from doing so. We have a sort of built in 'I cant do that' mentality for anything that is difficult. This negativity is supposed to be the reason we fail at complex tasks like, for me, learning Chinese. So what the recordings do when you are asleep is they supposedly reprogram your subconscious self image to be much more positive about the whole thing. They continuously repeat things like..."You want to learn Chinese. You enjoy studying and practicing it. You enjoy learning new words. When you hear or read a new word you focus on it, and it's meaning, and it enters your memory permanently. You recall the correct words quickly, easily and accurately. You speak it with fluency, confidence and the correct intonation." The idea then is that your subconscious mind is reprogrammed to view the task much more positively. Later, when you are awake, your actual behaviour changes because of these secret messages that go into your subconscious mind while you are sleeping

I was sceptical about it. The first reason for my scepticism is that this whole field of subliminal programming goes back to the 1950s and 1960s, and in particular one very famous experiment reported by a market researcher called James Vicary, who ran a company called the 'subliminal projection Company Inc'. Over a 6 week period Vicary manipulated the audiences attending a movie theatre in New Jersey in the USA. During the showing of a film he inserted two messages that were displayed for only 3/1000's of a second, a period that was too short for the movie goers to actually see the images. But the images were repeated every 5 seconds for the entire film. The messages said either "Eat Popcorn" or "Drink coca-cola". The experiment lasted 6 weeks and the films were shown to over 45,000 movie goers. To assess whether this had actually changed behaviour the scientist compared the sales of coca-cola and popcorn over the 6 weeks prior to the experiment with the sales during the experiment. Remarkably these subliminal messages had increased coca-cola sales by 18.1% and the sales of popcorn had gone up by a massive 57%.

The experiment created a real stir among the American public. If advertisers could make us buy products so easily where was it all to end. At the time TVs were becoming popular in peoples homes. Did this study mean that subliminal adverts could be beamed into their houses forcing them to rush out to the supermarkets to buy things. Was it ethically acceptable to tamper with people's minds in this way, without their consent, or without them even knowing it had happened? A typical newspaper report at the time read as follows "The subconscious mind is the most delicate part of the most delicate apparatus in the entire universe. It is not to be smudged, sullied, or twisted in order to boost the sales of popcorn." (from the Saturday review, October 5th

1957). But the fear was even greater than this. Maybe it was not just messages selling things, but perhaps subliminal messages could be sent to black Americans to encourage them to revolt against whites, or the absolute worst fear of all, messages could be transmitted to turn people into communists! Remember that this was at the very height of the US paranoia about the spread of communism in Russia and China. Fearing a backlash TV broadcast companies and theatre companies rapidly declared they would not accept subliminal adverts and such policies generally remain in place today. No actual laws banning subliminal adverts were passed, but the regulatory policies of the broadcast companies led much later to some interesting court cases in which the banning of subliminal messages was challenged as against the right to 'freedom of speech' by the people sending the messages.

They need not have worried however. Many studies attempted to repeat the work by Vicary, and they all failed to generate any significant effect. Why was it so hard to replicate this important work, when the original effects had been so large? One thing to think about is why the original work actually had any effect in the first place. Think about it. When you go to the movies, when do you buy your drinks and popcorn? You buy them BEFORE you see the film. Nobody buys these things as they leave the cinema. So why would exposure to the messages during the film increase their sales? It doesn't make sense. In fact in a TV interview five years after the work was published James Vicary admitted that the study was a fraud. At first he said that they had done the experiment and the results actually showed only a very minor change in the sales, and he had manipulated the data. However in 1992, an interview appeared with the manager of the movie theatre where the work was supposed to have been carried out, which said that the study had actually never been done at all. The whole thing was a fake. Despite this, and the fact all the subsequent studies also found no effect, there is a very strong belief still that it is possible to somehow access our subconscious minds by playing us messages either so short we can't consciously perceive them, or during our sleeping hours when we also are not consciously aware.

My second reason for being sceptical about the DVDs to teach me Chinese while I was sleeping was simply that if the method worked then everyone would be doing it. Everyone would be proficient at multiple languages and everyone would excel academically in whatever subjects they chose to be subliminally programmed to enjoy. It is such an attractive proposition to learn without taking out any time from our everyday lives that if it worked I would already know about it and be doing it in many other spheres of my life – and my university would be encouraging all its students to play the DVDs to make sure we turn out the most able and committed graduates. It seems our brains are unable to take in and use these messages to any significant advantage.

Another area where it seems our brains may also be unable to take in and process complex information is when we are babies or small infants. This is also a time when many parents invest in materials to try to teach their babies to read. There are many companies that produce such 'baby reading' materials, and they are incredibly popular both in China and in the west. Here is some of the text from a typical advert for these products on the internet "All babies are Einstein's when it comes to learning to read. Your baby can actually learn to read beginning at 3 months of age. Research shows that from this early age babies have to ability to learn languages, whether written, foreign of sign language with ease". The packs generally train your baby by presenting them with large words which are presented visually at the same time that the word is spoken. You can do the same thing a little cheaper by buying packs of 'flash cards' and saying the words yourself (something like the 'flash cards' I have to learn Chinese!). The testimonials of parents that have bought these packs are very positive. Of course the websites selling the products are not going to choose negative comments. However, if you look at the comments posted on independent evaluation sites, where parents can say whatever they like, the comments are also universally positive. Parents love these programs and make remarkable claims for how well their young babies are doing on them. The only negative comments I could find were things like the cards didn't stand up to the baby trying to eat them! Across 50 different products available I could not find a single comment from a parent who said something along the lines of "don't buy this, it doesn't work".

So based on my second reason why I didn't buy the 'learn Chinese while you are sleeping' DVD I have to admit that the teach your baby to read DVDs and flashcards seem a great buy. Maybe it really is possible to give your child a head start in their education by exposing them to such materials when they are still unable to even speak. Or are the companies who produce these programs just preying on the ambitions of parents who want to give their children an edge. The fact parents avidly buy them and report how good they are suggests not. However, some research just published in the Journal of Educational Psychology by scientists at the Steinhard's Department of teaching and learning at New York University, calls the effectiveness of these systems into question. The published study involved 117 infants aged between 9 and 18 months of age that were randomly assigned to one of 2 different groups. In the first group of 61 children, parents were given a baby media product, which included DVDs, word and picture flashcards, and flip books. The instructions to the parents were to use these materials every day over a seven-month period. The actual product they used was called 'your baby can read'. In the other group these materials were not provided. To test children's skills after these different experiences the children were invited to the laboratory where they undertook a series of tests. These included tests which examined the capacity to recognize letter names and letter sounds, words identified on sight, and their comprehension of what these words meant. They also monitored the size of the children's vocabulary. In addition to these standard measures the researchers also used a state-of-the art eye-tracking technology, which followed even the slightest eye movements, allowing them to monitor how the infants distributed their attention and how they shifted their gaze from one location to another when shown specific words and phrases. In total a battery of 14 tests were applied.

The results surprisingly showed absolutely no differences between the infants exposed to the baby reading media and the control group on 13 of the 14 assessments. The study leader Dr Susan Neuman concluded that this is probably not a problem with the particular product they tested but actually something common to all the products. The problem is that at this age babies brains are simply not wired up sufficiently for them to learn the complex task of reading. A positive outcome of the work, however, was that at least the exposure to the materials didn't do the children any harm! So parents had not spent their cash on something that damaged their babies, but they had still spent their cash without any tangible benefit. Given 13 of the 14 tests didn't show any difference between the groups you are probably wanting to know what the only test was that came up positive. The only positive result was actually the perceptions of the parents on the skills of their babies! Even though there was no actual difference between the babies that used the products and those that didn't, the parents who used the early reading product actually believed that their children were acquiring better reading skills than the parents who didn't use the product. And this result explains the anomaly of why, if the products do not work, do people keep buying them? The answer is that the people who do the actual buying strongly believe that they do work. But the reality is that there is no benefit in flashing too early.

Lazy? It could be genetic, and there could be a cure around the corner!

May 2014

Back in the UK I have a friend called Dave. Dave is a nice guy but he has one big fault. He is really lazy. You maybe know somebody like him. Dave always takes the elevator instead of the stairs. He always takes the car if he wants to go anywhere, and he never walks, not even if he is going somewhere really close by. If you go around to his apartment, then more often than not he is sitting on the sofa with the TV remote controller in one hand, and a large bag of potato chips in the other. In his kitchen are the unwashed dishes from several days of meals that he has not been bothered to wash up. In the western world he is what we call a 'couch potato'. The term couch potato really sums up Dave. His natural habitat is his sofa, and, especially when he doesn't have his shirt on, he really looks like a large uncooked potato. In fact, the term 'couch potato' is so popular that it has been trademarked in America where it was first invented. It has its origins in the 1960s and 70s among a group of people who were opposed to the increasingly popular activities of doing exercise and eating healthy diets. This reactionary group favoured instead sitting about watching the TV and eating junk food. Because the TV was known in some areas of America at the time as the 'boob tube' the group called themselves – the 'boob tubers'. Potatoes of course belong to a family of root vegetables that have enlarged roots called 'tubers'. A man called Tom Iacino saw this link in the words – 'boob tubers' being the people who liked sitting on the couch watching TV, and 'tubers' in the vegetable sense. He put the two together and the term 'couch potato' was born – on the 15th of July 1976 to be precise, during a phone conversation, if the story in the trademark application is to be believed.

Maybe the phrase wouldn't have caught on, but it was picked up by an American cartoonist called Bob Armstrong, and he created a cartoon strip called 'the couch potatoes' which featured the lives of two potatoes that lived on a couch perpetually watching TV. The 'couch potato' cartoons were really popular in the 1980s, and they appeared in hundreds of newspapers and magazines throughout America, until eventually the term became an accepted part of the language. Perfect to describe my friend Dave and the millions of others who are like him.   Armstrong and two colleagues even wrote a book called 'The couch potato guide to life' published in 1985. Although the 'boob tubers' and the 'couch potatoes' were created as a bit of a laugh, actually being a couch potato is no joke. The world health organisation has estimated that inactivity is one of the largest killers in the modern world. People who adopt a couch potato lifestyle have increased risk of developing obesity, diabetes, cardiovascular disease and even cancer. In other words the risks for developing the major killers in the modern world are all increased if you adopt a couch potato lifestyle. We have known this for over 5 years, since a whole series of epidemiological studies were published linking sedentary time to risks for various disorders including all cause mortality. A recent review study published in 2012 by Katzmarzyk and Lee from the Pennington Biomedical Research Center in the USA, which summarised data across several previous studies showed that for individuals who spent more than 4 hours per day watching TV, the risk of all cause mortality was 50% greater than those who watched only 2 hours per day. Dave actually knows that sitting around on the sofa all day isn't good for his health. In fact after a recent doctors visit it turned out that he had high blood pressure, high cholesterol and disturbed levels of his blood lipids (called dyslipidemia). Yet despite these warning signs that worse is on the horizon he continues to do it. He seems unable to motivate himself to break out of his lazy behaviour pattern. Medically Dave is not described as a couch potato but is instead said to have the 'metabolic syndrome'

Why my friend Dave can't break out of his lazy lifestyle is unknown, but one potential reason that has recently come to light, due to some research performed by Chinese Scientists led by Professor Wei Li, working at the Chinese Academy of Sciences, Institute of Genetics and Developmental biology, in Beijing, is that some people may be lazy and develop metabolic syndrome because of their genetics. The study published in the February issue of the scientific journal PLOS genetics primarily concerned a mouse called the *Ros* mouse. The *Ros* mouse is the murine equivalent of Dave. The mice sit around all day and move much less than their wild type relatives. Moreover, when they do move, they move more slowly, and they cover a shorter distance. They have lower energy expenditure because of their lower activity, and as a result the mice start to get obese around 8 weeks of age. By 6 months of age the *Ros* mice weigh on average 35 grams compared with only 25 grams in the wild type mice. Most of the weight difference seems to be due to large expansions of their fat tissue. Already by 12 weeks of age the mice have elevated blood lipids (by 61%) and elevated cholesterol levels (19%) and their fasting blood glucose levels are also 43% higher than the controls. Glucose tolerance tests, during which the animals are given a bolus of glucose, and then the levels in their blood are followed over the next 2-3 hours showed that the *Ros* mice had disturbed glucose homeostasis – a sort of early stage of diabetes. If they were humans they would be diagnosed as having metabolic syndrome.

The mutation in the *Ros* mouse is in a gene called SLC35D3, part of a large family of genes called solute carriers (hence SLC). In this case the gene name stands for Solute carrier family 35 member D3. The mutation involves a spontaneous insertion into the DNA which disrupts the function of the gene. This particular gene is expressed in platelets in the blood, and also in very restricted areas of the brain called the substantia nigra, striatum and the olfactory bulb. The substantia nigra and striatum are small areas of the brain that are part of the dopamine system. They appear to have a key role in the control of reward and addiction, but also critically they are involved in the regulation of movement. Parkinson's disease, which is common in old age, is characterised by disturbances of movement such as tremor in the hands, and is caused in part by the selective death of neurons in the substantia nigra and striatum. The fact the *Ros* mutant mice have reduced activity is therefore consistent with this gene playing a role in the dopamine system in these two restricted brain regions. In fact within these areas the gene is expressed only in one type of neuronal cell which expresses dopamine receptor 1, but not cells expressing dopamine receptor 2. Li and colleagues therefore argued that SLC35D3 might be playing a role in dopamine signalling, and thereby affecting movement levels via regulation of the dopamine 1 receptor (D1R).

In fact, in the *Ros* mouse there appeared to be a different cellular distribution of the dopamine 1 receptor. The *Ros* mice had more D1R inside their cells but rather less (about 17% lower) on their cell surfaces. When proteins such as the dopamine receptors are manufactured in cells they are produced on ribosomes that are located in a subcellular structure called endoplasmic reticulum (ER). These proteins are then trafficked to the cell surface via endosomes and a structure called the Golgi apparatus. SLC35D3 was located primarily on the endoplasmic reticulum and early endosomes, but not in late endosomes and the Golgi apparatus. This suggested that its function in the cell was connected either to protein synthesis or its early export from the ER. Consistent with this model *Ros* mutant mice had elevated levels of D1R in their endoplasmic reticulum, and this failure to export the synthesised receptor was the primary reason these mice had lower D1R levels on their neuronal cell surfaces. If the lower levels of D1R on the neuronal cell surfaces was the main cause of the effect of the *Ros* mutation of SLC35D3 then Li and colleagues argued that perhaps the effect could be counteracted by putting in a specific agonist of the D1R to stimulate the receptors that were present. Injecting such an agonist every day for 12 days caused a 13% drop in body weight of the *Ros* mutants (compared to only 7% in the wild type individuals) and rescued their elevated levels of cholesterol, lipids and glucose. So it appears that in these mice at least a simple mutation in a single gene is enough to impact the trafficking of dopamine 1 receptors to the neuronal surfaces, and that this makes them less active. Their lowered activity then leads to all the same sorts of issues that follow on when humans are inactive.

Obviously this is just a mutant mouse. What relevance does it have to humans? Is it really credible that the same genetic mutation causes laziness in people? The answer seems to be "sometimes". One of the real strengths of the paper by Li and colleagues was that they not only worked on mice but they also screened a population of 363 Chinese people with symptoms of the metabolic syndrome (low physical activity, obesity, dyslipidemia and elevated glucose) living in Qingdao and Nanjing, and compared them with 217 healthy people. Among the 363 affected people two were found with mutations in SLC35D3, but none of the healthy people had such mutations. The mutations were not the same as each other, and they also differed from the mutation in the mice. However, assuming that in these 2 individuals this mutation was responsible for their metabolic syndrome suggests about a 1/200 chance that being a couch potato can be traced to mutations in this one gene. In fact a much bigger study of Caucasians in America called the Framlington heart study also found that there was a genetic mutation close-by to the SLC35D3 gene that was also linked to the same symptoms of metabolic syndrome, so this may also be true in Caucasians as well as Chinese people. What about the remaining 199/200 – are they just lazy. Probably not. Indeed last year another group working in the USA also found an association between a particular gene and the metabolic syndrome, although in that case the exact functions of the gene were not worked out in the same way as in the comprehensive study by Li and colleagues. Probably many other genes that cause metabolic syndrome are out there waiting to be discovered. This study is the first rather than the last step in the process.

Perhaps knowing that his behaviour is possibly driven by a genetic mutation will give some comfort to Dave that his behaviour is not because of some personality defect. However, what should give him even more comfort are the experiments in the mice where they were given a dopamine receptor 1 agonist which reversed the symptoms and made the mice thinner and less metabolically affected. It is possible to imagine in the future that people who suffer from metabolic syndrome might be screened to see if they have this particular mutation (or others) and if so they could be given a drug to reverse the effects. This study therefore brings us one step closer to a future where our genetics is used to tailor the treatment to fit the genetic abnormality underlying whatever problem we have – the so-called era of personalised medicine.

Crossing the boundary

June 2014

Sixty years ago this month (June) a young British scientist was found dead, 16 days before his 42 birthday, in his small apartment in the English city of Manchester. The official inquest decided it was suicide. The man was found dead by his cleaner with a half eaten apple beside him, and the autopsy report subsequently showed that the cause of death was cyanide poisoning. Following his death several people, including the scientists' mother, questioned the verdict of suicide, and instead suggested that it was more likely due to an accident, because he had set up equipment in his home to gold electroplate spoons – a process that requires potassium cyanide to dissolve the gold. In support of this alternative idea there was no suicide note, and among his possessions at the time of his death there was a list of tasks that he intended to complete after the forthcoming holiday weekend! Whatever the cause, the strange death brought to an end the life of a remarkable man. His name was Alan Turing, and he is widely regarded as one of the most brilliant scientists of all time, the father of modern computing and the study of artificial intelligence.

Born in London just before the first world war, his father worked for the UK government mostly in India. Turing showed many early signs of brilliance, and eventually went to the University of Cambridge to study maths for his undergraduate studies, where he obtained a first class honours degree. Following his degree, he became a research fellow at Kings college in Cambridge and developed what were to become known as Turing machines. These are hypothetical machines that manipulate symbols printed in paper tape. This machine can be adapted to simulate the process of computing, and from this was born the idea of a computer algorithm. Turing showed that, in theory, it would be possible to develop a machine that was capable of computing anything, that was capable of being computed, as long as an algorithm could be written to instruct the process of calculation. This is the central concept underpinning the mechanism by which the central processing unit of a computer works. After this early success in maths and computing he went to Princeton University in America. Princeton has been the home of many famous scientists including Einstein, and 22 other Nobel prize winners, and in addition two of its graduates became American presidents. Yet in the *Princeton alumni weekly* it was Turing who was listed as the second most important person ever to have attended the University (after the US president James Madison who is credited with being the architect of the American constitution). Turing obtained his PhD from Princeton in 1938 which was based on pure maths and some early work on codes.

On his return to the UK he started to work part time at the government research centre for the development of ciphers and codes. A year later, the second world war started, and Turing's work became increasingly devoted to deciphering messages sent by the Germans. He made several important contributions to the mathematical basis of code making and code breaking during this period that were so important and fundamental that they were not allowed to be published by the UK government until 2012, to coincide with celebrations of the 100th anniversary of his birth. These were amazing achievements, but it is clear that Turing was no ordinary person. In fact not only was he a brilliant scientist but he was also an accomplished athlete, specialising in long distance running. He would occasionally run the 40 miles from Bletchly park where he worked, into central London, for high level government meetings! It was immediately after the war ended that Turing published a paper which described the first detailed design for a computer with an on board stored program: called the Automatic Computing Engine or ACE. A pilot version of this machine was built by the UK government, and executed its first program in May 1950. This event is regarded by many as the start of the computer age.

In 1948 Turing went to the University of Manchester in the north of England and it is there that three remarkable things took place. The first I have already told you about, and that is the strange nature of his death. The second thing was related to his private life. Turing was in fact a homosexual. During his time in Manchester he met another man called Arnold Murray. During January 1953 Turing's house was burgled and Turing reported it to the police. During the police investigation it emerged that Turing was in a relationship with Murray, and he was subsequently tried and convicted of gross indecency, because at the time homosexuality was illegal in the UK. Turing was given an option of going to prison, or of being chemically castrated, and he chose the latter. He was injected with a synthetic form of the female hormone oestrogen for a year, which made him impotent, and it also caused him to develop breasts. Some have speculated that his 'suicide' a year later was related to his treatment by the authorities for his homosexuality, but there is no evidence that he was psychologically negatively affected by this treatment. In 2009, a petition was started in the UK to get the UK government to apologise for the way that Turing had been treated. Under public pressure a public apology was made by the government in late 2009, this was followed by a further petition to have his conviction quashed. At the end of last year the Queen of the United Kingdom issued a royal pardon for him. This was in fact only the fourth royal pardon to be issued in the last 60 years. The remarkable thing about this pardon is that nobody has ever claimed that Turing was innocent. So this was the first time ever that a Royal pardon has been given to someone who never denied their guilt on the issue at hand.

The third remarkable thing to happen in Manchester in the years he spent there after the second world war was Turing's work on artificial intelligence. This led to what has become known as the 'Turing test'. Turing wanted to define a standard by which a machine could be said to be 'thinking'. This problem has in fact vexed philosophers for many centuries. In attempting to devise a test to answer the question 'can a machine think?' Turing actually devised a rather different test. Turing's test is based on the ability of a machine to deceive a human in conversation that it is not a machine. That is, if a person converses with the machine and is unable to tell if the machine is a human or not, then the machine can be said to be thinking. Turing said that to pass the test a machine would need to be able to deceive the average interrogator more than 30% of the time following 5 minutes of questioning. Although it has been widely criticised the Turing test remains today as one of the major discrimination tests for whether artificial intelligence has advanced to the point of really being able to think. Turing thought that the test would be passed by 2000. He was wrong. By 2000 no artificial intelligence had passed the test. The futurist Ray Kurzweil predicted in 1990 that with the rate at which computer technology was expanding there would be artificial intelligence machines able to pass the Turing test by 2020. However in 2005 he revised this estimate to 2029. This prediction led to the 'long bet project' which is a 20,000 US dollar bet between Kurzweil who believes that a machine will pass the test by 2029 and a colleague (Mitch Kapor) who believes it will not.

If recent reports prove to be true then Kuzweil has

won the bet, because on the 60th anniversary of Turings death, on the 7th June 2014, it was reported that a computer had passed the Turing test. The computer is called 'Eugene Goostman' and it is a simulation of a 13 year old boy. The program was created by collaboration between a Russian born scientist called Vladimir Veselov, who currently lives in the USA and a Ukrainian called Eugene Demchenko, who lives in Russia. The machine was tested along with 4 other computers in an event performed at the Royal Society in London (the UK equivalent to the Chinese national academy), timed to coincide with the anniversary of Turing's death. Eugene Goostman managed to convince 33% of the people it interacted with that it was actually a human, and so came over the 30% limit set by Turing. The organisers of the competition in London have been heralding this as a true landmark in the development of artificial intelligence. The point at which artificial intelligence machines first crossed the boundary set by the grandfather of computing over 60 years ago, and started to 'think'.

Time will tell, however, if this is really the landmark that some are claiming. The main problem is that the computer does not claim to be able to simulate a human, but a 13 year old boy. They are not quite the same thing. Anyone who has raised male children will know that producing a simulation of a 13 year old boy is a rather more simple task than simulating a grown adult, because their range of reaction to things is at best described as 'limited'. Most real 13 year old boys would struggle to hold down a 5 minute typed conversation with an adult, and so a computer that responded to questions with a variety of responses like 'yea', 'I guess so', 'I dunno' and 'whatever', could probably go a long way to duping 33% of people it interacted with into thinking it really was a 13 year old boy. It is revealing that the organisers of the competition have not actually released the transcripts of the conversations that 'Eugene' had with people. In addition I would be interested to know what the results would be if you substituted a series of real 13 year old boys in place of the computer. How many people would classify these boys as robots? Perhaps only 33% of real 13 year old boys would be identified as human by the same test! In fact we already have "reverse Turing tests" which aim to identify if someone is a real person or not. The most common of these is the so called CAPTCHA test which many web sites use. This is when the computer asks you to type back a series of distorted letters or to write the answer to a simple arithmetic problem. In these tests the computer is trying to establish if you really are human. So in the case of the Turing test what we need as a control comparison is the parallel results of a reverse Turing test using actual humans trying to fool people as to whether they are human or not! The unstated assumption in the 'Turing test' is that humans would always pass the test. I am not so sure they would, particularly if they were male and 13 years old.

I suspect that 'Eugene Goostman' will not actually

turn out to be quite the 'crossing of a boundary' as it initially seems to be. However, if it is, or is not, it is clear what direction the wind is blowing. If the boundary was not crossed on the 7th of June 2014, it will be crossed soon. I don't think we will have to wait until 2029. The implications when it is eventually crossed will be enormous. This article was written by a human. How long will it be before you cannot tell?

Youthful blood

July 2014

In 1849 the ruins were discovered of a library from the ancient country of Sumaria, located in modern day Iraq. The library dated back to 700 BC, and contained thousands of stone tablets covered in a strange form of writing – called 'cuniform' writing. Once they were translated it became clear that some of the tablets tell a story of a king called Gilgamesh who had ruled in the same area about 1000 years earlier (1800 BC). The 'epic of Gilgamesh' as it has become known is one of the first ever works of literature. The theme of the story is the quest for immortality, and it is in two parts. The first part of the story is about King Gilgamesh making fiends with a young wild man called Enkidu. Gilgamesh and Enkidu go on a series of great adventures with each other, but on one adventure Enkidu cuts down a forest and this behaviour angers the gods who sentence Enkidu to death and kill him in revenge for his actions.

Gilgamesh is distraught. Not only has he lost his friend but he has realised that one day he too will die, like all people do. In the second half of the story, he sets out on a quest to find the secret of immortality. Gilgamesh travels to the ends of the earth until eventually he meets a half man half god figure called

Utnapishtim, who tells him that the quest for immortality by man is fruitless, because only the gods can live forever. When god made man, he tells Gilgamesh, he gave him death to share as his partner forever. However, it isn't all bad news, what humans can aspire to is a long and healthy life, not plagued by the problems of old age and disease. Utnapishtim says that this can be achieved by taking a medicine which he gives to Gilgamesh in a bottle. Overjoyed, Gilgamesh decides to return home to his kingdom to share the great news and the formula of the medicine with his entire kingdom. On the way, however, he falls asleep and while he is sleeping the bottle is stolen by a snake, and the formula is lost forever. It is interesting that one of the first ever written stories should address the theme of longevity and preventing death. As soon as people were able to think this must have been one of their abiding questions. Why do we get old? why do we die? And what can we do to delay the ageing process and our inevitable death?

Four thousand years after Gilgamesh ruled in Sumaria, these questions are still uppermost in our minds. At the turn of the millennium the journal Science published the 100 most important outstanding questions for scientists to tackle during the next 100 years. Top of the list was why do we age and die? We are still engaged in the same quest as we were 4000 years ago. No one has yet found the formula that Utnapishtim gave to Gilgamesh. However, even if 4000 years ago it was just a story, the modern day quest to find things that retard the process of ageing is very real, and scientists are making some big breakthroughs in understanding the ageing process and what we might do to give ourselves some extra time, or at least improve the quality of the time we have.

In May this year three papers were published in two high profile journals (Science and Nature medicine) which suggested that there may be a secret ingredient in our blood that keeps us young. The papers all build on observations that have used a conceptually simple, but technically complex method to see if our blood contains age related factors. The technique is called parabiosis. It has been used for over 150 years to study biology and it involves joining two individuals together surgically so that their blood systems mingle together. Normally, when studies like this were first made the animals (most often rats) that were combined together were siblings, but an extension of the method was to combine together non-siblings of very different ages. Over 60 years ago Clive McCay a researcher into ageing at Cornell University performed some experiments on rats using this approach. He stitched together old (300 day old) rats and a young ones (just 60 days old) and he found obvious improvements in the older individual matched by reductions in performance of the younger one. This strongly suggests that our blood contains chemicals that help to maintain our youthfulness, and perhaps also negative things that promote our old age. Unfortunately for McCay the technology at the time was unable to identify these factors and the use of parabiosis as a tool to study ageing fell out of fashion.

About ten years ago, however, scientists at Stanford University in California led by Thomas Rando started to use the method of young-old parabiosis again. Two of the three papers published in May used exactly this approach and looked at what happens in the brains of the co-joined rats. One study, in Nature medicine, was led by Tony Wyss-Coray from Stanford. In detail they studied the hippocampus, the area of the brain that is important for memory formation. During parabiosis they found that the neurons in the hippocampus of the old mice made more connections: their dendritic spine density improved and so did their synaptic plasticity. This would suggest their memory functions should be improved so they extended the work by injecting old mice with plasma from the blood of young mice three times a week, and found that such mice were indeed better at memory tasks, like learning mazes, than untreated control individuals. For example, the maze navigation error rates of the injected mice were 25% lower than the controls.

This raises some really interesting prospects if the same effects are found in humans. Indeed the researchers are already talking about human trials with patients who have Alzheimer's disease to see if blood transfusions from young people will help reverse their disorder, and a company has been set up in the USA to help promote this approach. One attractive aspect is that since plasma transfusions are already given routinely to patients this approach would not need a new approval by the Federal drug administration so could be move from trials to routine practice very rapidly. As interesting as this is, however, transfusing blood from young to old people has obvious limitations and potential problems. It is unclear, for example, how frequently it would be necessary to receive a transfusion. If to remain youthful one would need a transfusion of say a litre of blood three times a week (like the mice in Nature medicine got), then there is a clear mismatch between the supply and demand, since routine practice is not to collect more than 0.5 litres of blood every 2 months, to give the lost red cells time to replenish in the donor. For each old recipient therefore you would need about fifty young donors.   In addition, demands for blood for other purposes like operations are increasing. If a large population of old people started to require significant amounts of 'youth giving' blood, then demand is clearly going to outstrip supply – potentially harming donations for surgical procedures. You might be youthful again, but if you then fell off your skateboard and needed surgery there might be no blood available to get you through it. In addition, this assumes that young people will willingly donate blood. For operations and to cure people of Alzheimer's, disease, maybe, but to provide older people with the fountain of youth, probably not. At least not for free.

One might imagine therefore a clear market potential where young people in the future recognise they have a resource flowing in their veins that older wealthy people might be willing to pay for. Inevitably one can envisage that this will lead to all sorts of difficulties. For a start there will not be enough young blood available for the demands of all the old people, particularly in China where the ratio of old to young people is high because of the operation of the one child policy. The resource will be scarce, the price will be high, and so this will be something only rich people will be able to afford. We can imagine, for example, future 'blood dealers' acting as intermediaries in this transaction, operating 'shops' where they extract the blood from young people, at a given price, and then sell it on at a profit to wealthy old people. I can see the 'shops' now in affluent suburbs like Sanlitun in Beijing (if you are in the US the equivalent would be Manhattan in New York, and if you are in the UK think about Chelsea in London). There nestled among the designer clothing shops, would be a rejuvenation center, where you could go in and select from a 'blood menu' a litre of blood from a 25 year old, or if you are really rich a 16 year old, and they infuse it in a back room somewhere, while you read a magazine, drink a coffee and have a foot massage, while in a poorer part of town (or a poorer country!) some young person is getting the same stuff drained out of them. You can see the obvious problems. If a young person goes around several shops donating blood, who will know when they have given too much for their own good – we may end up literally draining life from the underprivileged young, to extend that of the well-heeled old. Sooner or later this will end up in a scandal where some dodgy dealer is caught selling prosperous old people blood that isn't actually from a young person (or from someone who is even older!). Plus there are all sorts of other possibilities from this sort of transfusion, chief among which is the fact that blood may not only carry the secret of youth but also the vectors of disease. You may have thought what you were buying was eternal youth, but what you actually got was HIV or hepatitis B.

Much better would be if one could identify what the magic ingredient in the blood is, and then manufacture that, without the need to collect blood from young people and infuse it into old people at all. One could simply inject the compound directly. This is what the other 2 papers published in May concerned – the identification of the magic ingredient. Screening old and young mouse blood using modern proteomic methods to identify differentially expressed proteins was actually done over a year ago by a team of scientists from Harvard university in the USA, led by scientist Amy Wagers, who was originally a post-doc in Rando's group at Stanford, where the young-old parabiosis method was revived. The list of targets was gradually narrowed down to a shortlist of around 13 proteins. One of these is called growth differentiating factor 11 (GDF11) a member of the transforming growth factor beta (TGFbeta) family.

GDF11 seems to act by regulating stem cell activity. As rodents get older their heart muscles get thicker and stiffer. Last year it was shown, in a paper in the journal *Cell*, that injecting GDF11 can reverse these signs of aging of the rodent heart. The new papers show how this exact same protein also has positive impacts on skeletal muscle and the brain ageing. In particular, one paper in *Science*, by the Wagers group, shows it has positive impacts on both muscle strength and running performance. A second paper in *Science*, from another group at Harvard, led by Lee Rubin, also used parabiosis to show young blood encourages the growth of new blood vessels and increases numbers of progenitor stem cells in the brain, and leads to an increase in neurons in the olfactory bulb in the older member of the parabiosis pair, but conversely the younger member of the parabiosis pair had reduced numbers of neural stem cells. Interestingly, these positive impacts on the old mice could be mimicked by daily injections of GDF11.

So is that it? Problem solved. We just need to inject GDF11 and we can all, or at least the rich can, stay young forever. Not quite. There are a few problems – apart from the fact this is work in rodents at the moment, and we don't know if it will work in humans. It seems unlikely that GDF11 does everything. There may be multiple factors involved that are necessary to get the full range of benefits that occur when transfusing plasma. Then, there is the observation that in parabiosis the young mice got worse, suggesting the accumulation of factors that have negative effects as we get older. So a complete rejuvenating treatment would need not only infusion of the positive youthful compound(s) but removal of the bad stuff as well. Finally, even if it did turn out that GDF11 is the only important rejuvenating compound required, the amount of protein needed would be huge, and since it is a protein it would need injected rather than being taken as a pill. However, the prospect of identifying the pathway by which GDF11 exerts its effects opens up the exciting possibility to derive non-protein pharmaceuticals that have the same effect – and then the metaphorical search for the secret formula given by Utnapishtim to Gilgamesh really would be over. Unless a snake was to steal it again!

Tibetans have ancient genes for running in thin air

August 2014

I have spent almost my whole life near to sea level. In the UK I live in a small town that was once a fishing village and is located right on the sea shore. In China I live in Beijing, which is only at an altitude of 44m above sea level. The first time I went to a really high altitude was about 5 years ago. I was part of a team that went to a site in Qinghai province to study the plateau pika – a small lagomorph that lives on the Tibetan plateau. The study site that we went to was at 4000m. Before we left to go to the site I was a little apprehensive about the effects that such high altitude might have. At 4000m the air only has about half the pressure that it does at sea level. So although the percentage oxygen is still 21%, that actual amount of air molecules that come into your lungs on each breath is only half that at sea level. That means the amount of oxygen available to fuel your metabolism is similarly halved. The drive south to the study site from Xining took us about 10 hours. It involved going over a series of progressively higher and higher mountain passes. As we climbed the twisting road up the sides of the mountains, I fully expected to start to have problems breathing and then eventually to get light headed and maybe even pass out. However, I was surprised to find that there were no effects whatsoever. When we stopped at the top of one of the passes we got out of the car to look at the view and take some photographs. I could actually have been anywhere. There was no indication at all that the air was considerably thinner than what I was normally used to breathing.

When we arrived in the town next to the field site we went to check into the hotel. We had a lot of bags with us and while I did all the form filling that Westerners need to do when checking into hotels there was a steady stream of people bringing all our bags from the van into the hotel lobby. Finally, we got checked in. My room was on the fourth floor. I had two large suitcases that were filled with heavy gear and weighed about 35 kg each. "That girl will help you with your cases" the hotel receptionist said. I turned around and behind me was a small girl whose bright red cheeks and round face immediately identified her as a local Tibetan. She picked up one of the cases, which seemed almost as large as she was, swung it onto her back and proceeded to the stairs. I grabbed the other case and followed her. By the fifth step I was panting for breath. A couple of steps later sweat started to drip from on my forehead. At the tenth step I was already a little dizzy and by the top of the first flight of stairs I dropped the case and collapsed on top of it. The girl looked back and laughed a little before running up the remaining flights. I felt sick and had to sit down for 5 minutes with my head between my knees to prevent me from fainting until I had recovered. My previous whole life at sea level had finally caught up with me. As soon as I started to do something strenuous my lungs simply couldn't drag enough oxygen out of the thin air to match my energy demands. I had probably switched to anerobic metabolism to keep going, but eventually I had to pack that in as the lactic acid built up in my muscles. As I sat there thinking I was going to throw up on the hotel carpet, I remember wondering how the young girl disappearing in the distance was able to carry such a heavy weight with apparent ease in such a thin atmosphere. The difference between her performance and mine was remarkable.

The obvious answer is that it was in her genetics. There have been lots of studies into the adaptations that people who live at high altitude have to help them get over the low oxygen levels. In South America, in the high Andes, indigenous people have elevated the amount of red cells in their blood to compensate for the reduced loading of oxygen onto their haemoglobin molecules. In fact when people from sea level travel to high altitude they respond to the low oxygen pressure in very much the same way. The increase in haemoglobin is stimulated by switching on a gene called Erythropoietin or EPO. Athletes have known about this effect for a long time and have used training at high altitude as a method to switch on the EPO gene, and enhance their levels of haemoglobin and hence oxygen delivery to their cells. When the same athletes return to sea level their performance is enhanced for a while because of their greater haemoglobin levels increase their capacity to deliver oxygen to their muscles.

I once met an international middle distance runner at a sports science conference and he said that the month following a training session at high altitude in Kenya was one of the most exhilarating and depressing periods of his life. Exhilarating because when he first came back from the high altitude training he ran the fastest 1500m race he had ever run, but depressing because over the next month his times progressively got slower and slower and returned to his pre-training levels. He was depressed because he knew that by staying at low altitude he would never again repeat the personal best he had set when he came back from high altitude training.

Once the stimulation of the EPO gene had been identified as the molecular mechanism underpinning this effect, then it was possible to get exactly the same impact as high altitude training by staying at low altitude and injecting EPO. Synthetic EPO is produced as a medicine to increase haemoglobin levels in people who have anemia. Rather than spend thousands of pounds travelling to train at altitude, many athletes got their enhanced performance in the form of an injection – altitude in a needle. Probably because of the impact on endurance performance this habit was particularly prevalent in the sports of long-distance cycling and running. EPO is one of the drugs that the disgraced cyclist

Lance Armstrong used to enhance his performance when he was winning the Tour de France cycling race. Sporting authorities banned the use of injected EPO in the 1990s, and tests for it were first implemented at the Olympic games in 2000, held in Sydney, Australia. If EPO is a natural substance then you may be wondering how is that possible? The tests were based on the fact that synthetic EPO has a slightly different structure than natural EPO. So it was possible to tell if a person had got high EPO levels from going to train at altitude, or had sat at home injecting themselves. It seems strange however that training at altitude, which we now know works via stimulating EPO, is completely legal in sport, but staying and training at low altitude, and injecting the same compound that causes the altitude effect, is deemed to be cheating. Indeed, it is now possible to simulate high altitude in low pressure chambers and currently these are also not banned by the world anti-doping agency. The reason for the ban on injecting EPO is because its effects are not all positive. Increasing the number of blood cells and hence heamoglobin levels is good for loading up oxygen into the blood and hence for enhancing athletic performance – but it makes the blood much thicker and that causes problems, because it is then harder to pump it around the body. That leads to high blood pressure, and when the person is at rest an increased probability that the blood will simply clot in the veins – or cause a stroke. There is a clear downside to that transient elevated performance, apart from the cycle of elation and depression it may induce.

So what about the young Tibetan girl carrying my ridiculously heavy case with such ease? Did she have high EPO levels? Probably not. However, more likely is that she had variants in other genes that allowed her to effectively use the low levels of oxygen that left me gasping for breath at the top of the first flight of stairs. What are these genes for such amazing low oxygen athletic performance? Four years ago a team of Chinese and American researchers identified variants of several genes that are involved in oxygen processing. These variants were found in many Tibetan highlanders. The paper was published in the prestigious journal *Science*. The gene variants were actually found using two methods. First, they scanned DNA registries for genes that might be involved in regulating the levels of oxygen in the blood and identified 247 candidate genes. Then they analyzed the DNA for these 247 genes in 31 Tibetans, 45 Han Chinese, and 45 Japanese lowlanders. It is possible from the genomic regions surrounding a particular gene to recognise if that gene has been under strong natural selection. Using this so-called 'selective sweep' method they were able to identify relatively new gene variants that had swept through highland Tibetans, but not the Chinese or Japanese people from low altitude. This approach identified a cluster of about 10 genes that appear to be functionally linked to efficient use of oxygen that have been under intense selection since the Tibetan population separated from a common ancestor with the Han Chinese population and spread into the high Himalayas around 2750 to 5500 years ago. One of these genes was called EPAS1, and it is related to the EPO gene, in that it appears to be linked to the production of haemoglobin. A variant of this gene appeared to have spread very rapidly though the Tibetan population. The initial calculations suggested that it had spread through 87% of high-altitude Tibetans in just 3000 years. This was the fastest genetic sweep that had ever been observed in humans.

Now, an international team of researchers has sequenced the *EPAS1* gene in 40 Tibetans and 40 Han Chinese and their work has also been published this month (July) in *Nature*. They found that all the Tibetans in their sample and only two of the 40 Han Chinese had a distinctive segment of the *EPAS1* gene. When they searched the catalog of genomes from people around the world, in the 1000 Genomes Project, they could not find the same sequence variant in a single other person. Because the gene variant was found in two of the Han Chinese that had been sequenced it seemed likely that the gene predated the split in the Han/Tibetan populations. To establish if this was the case the team compared the gene variant with DNA sequences from ancient humans, including Neandertals and a Denisovan,. The Neanderthals were a species of hominids found in Europe until about 25,000 years ago. The Tibetan gene variant in EPAS1 was not present in the Neanderthal genome. The Denisovans were a separate species of hominid that lived in the Altai mountains of Siberia. They were identified initially from the mitochondrial DNA extracted from the finger bone of a young girl found in the Denisova cave close to the border with China and Mongolia (hence their name). This analysis showed that the most recent common ancestor of the Denisovans, Neanderthals and modern humans lived over 1 million years ago confirming the Denisovans were indeed a separate species of the genus Homo. The finger was dated to around 40,000 years ago, but it isn't entirely clear when they went extinct.

Although the Denisovans are no longer with us their genetic legacy lives on because it seems they interbred widely with modern humans that they encountered. Consequently Denisovan genes are found in many modern human populations. About 2-4% of the genome of modern Papua New Guineans and Melanesian islanders is Denisovan. And what is amazing is that the same gene variant in the EPAS1 gene that was found in the Tibetans appears to also be a Denisovan gene! By comparing the full *EPAS1* gene between populations around the world the team confirmed that the Tibetans' inherited the entire gene from Denisovans in the past 40,000 years or so. The alternative explanation that there was a common ancestor from around 400,000 years ago, that carried this variant and passed it on to both Denisovans and modern humans could be eliminated because such a large segment of DNA would have accumulated many mutations and the Tibetans' and Denisovans' versions of the gene wouldn't have matched as closely as they do.

How did this gene end up in the Han Chinese and the Tibetans? The only plausible explanation was that the ancestors of both Tibetans and Han Chinese got the gene by mating with Denisovans. This is consistent wth the fact that some Han Chinese and mainland Asians retain a low level of Denisovan ancestry (about 0.2% to 2%). Most Han Chinese lost the Denisovans' version of the *EPAS1* gene, presumably because it conferred no advantage at low altitude and high oxygen levels. The gene was retained in Tibetans and it spread rapidly through the population because of the advantages it confers for living at high altitude. A few Han Chinese—perhaps 1% to 2%—still carry the Denisovan version of the *EPAS1* gene today. It isn't clear however whether this is a remnant from ancient history, or due to more recent inter-breeding between Han Chinese and Tibetans, since these populations are clearly not completely reproductively isolated. The bottom line however is that the young Tibetan girl in my hotel left me gasping for breath on the stairway while she ran up the remaining flights, likely because of a genetic advantage she inherited from an ancient population of hominids many thousands of years before. Which got me to thinking? If it is illegal in sporting competition to inject a natural gene product like EPO, then why is it not illegal to have a genetic variant that influences the same gene (or a related gene like

EPAS1) if that also gives you a performance advantage? How long will it be in the future that having a genetic advantage in a sport will be considered cheating? This may seem fanciful, but think about it. If it isn't made illegal, then how long will it be until someone engineers a future Olympic gold athlete by gene therapy?

Stimulating memories

September 2014

Let's play a little game. I am going to give you a 7 digit number. I am then going to ask you to do something, and once you have finished doing it I want you, without looking at the page, to try and remember what the number was. Ok let's start. Read the number that follows and then close your eyes and while they are closed say the name of your mother and your father and then try to recall the number. The number is **9462381.** Do it now. How did you perform? Most people given this task can get the number exactly correct, or at most make just one or two errors.

Ok, lets play again. This time it's a bit harder. This time I will give you the number, and then when you close your eyes I want you to subtract the number 7 repeatedly from the number 1001. That is 1001…994….987….980…..973…etc Do this until the number you have left is less than 900. Then immediately try to remember the seven digit number without looking at the page. The seven digit number this time is **7421184**. Close your eyes and start subtracting now. How did you do? Not as well I bet. In fact most people can't recall more than the first digit. Don't look up the page – can you remember the first 7 digit number I gave you? Probably not at all. Maybe not even the first digit.

Now tell me what your phone number is. Your phone number is probably more than 7 digits, yet you had no problem recalling it instantly. The reason for the differences in what we can recall, and what we cant recall, is because we have two different types of memory. The first is called our working memory. The second is called long-term memory. Working memory only lasts about 30 seconds. The reason you could recall the first number when I told you to close your eyes and say the names of your mother and father was that it only takes about 5 seconds to do that, and so the number was still in your working memory when you came to recall it. On the second task, however, unless you are particularly fast at doing subtractions, the task I gave you lasted much longer than 30 seconds. It actually doesn't matter what the task you have to do in the interval. The purpose of the subtraction task was just to stop you from saying the number over and over to keep it actively renewed in the working memory. Since that period was longer than 30 seconds when it came to remembering the number you couldn't, because it had gone from the working memory – and of course so had the first number, which is why you couldn't recall that either.

Your phone number, however, isn't in working memory. It is in long-term memory and it can be recalled over much longer periods. When I was a child the phone number of my best friends house was 674550. I have not phoned that number for over 35 years and yet when I came to remember it just now, to write it down, I did so instantly without an error. Some long term memories can be really long term, seemingly impossible to erase. But long term memories are not indelible. Sometimes that is a good thing. There are some memories that it would be too painful to remember forever. Forgetting is an important thing to be able to do. It has been shown, for example, that women forget the pain of childbirth much more rapidly than the equivalent pain associated with a traumatic accident, like a car crash. Clearly, if one remembered in excruciating detail the pain of childbirth then women would never have a second child! Forgetting in this situation is adaptive. But the problem is we often forget things that really we do want to remember: like the answers to questions in exams. What makes the difference between something in my long term memory that is there forever, like my childhood friends phone number, and things that I once knew, but now have forgotten, like the first integral of $\sin(x)$, or the date that Abraham Lincoln was born. If we knew how this system worked, then perhaps we would be able to learn and remember everything we wanted to: perfectly, and forever! There is clearly a market for this need, if the number of people selling supposed 'memory enhancing drugs' around exam times at universities in the UK is anything to go by.

How does the system work? How do the things in short-term memory get transferred into long term memory, and what affects the permanence of that transfer process? By way of explanation I am going to take you on a little diversion. In 1953, in the USA, there was a young man called Henry Molaison. Henry had been plagued all his life by epilepsy. He had epileptic fits every day, and these fits were becoming more and more serious in duration and intensity. It was becoming difficult to control his problem, even with large doses of medication. Something radical needed to be done, and so Henry was enrolled into an experimental program to see if these epileptic fits could be cured by removing part of his brain. On Tuesday August 25th 1953 Henry was admitted for surgery. He was 27 years old. In one sense the surgery on Henry's brain was a great success. The epileptic seizures that had so blighted his life up to that point were dramatically reduced. But the cure for his epilepsy came at a terrible price. Immediately, after the operation Henry was taken to a recovery room where he was watched every 15 minutes to make sure there were no complications resulting from the surgery. Eventually, he went back to his hospital room, where he was reunited with his parents. Over the next few days it was clear that Henry had been negatively affected by the surgery. He would forget the names of the nurses who tended him. He forgot where the bathroom was, even if he had visited it just a few hours previously. He repeated himself endlessly. Quite often patients recovering from brain surgery have a period of confusion, but in Henry's case the problem persisted, and eventually it became clear that while the surgery had cured his epilepsy, it had severely disrupted his memory processes.

Between 1955 and 2008 when he died, Henry Molaison became one of the most studied humans on the planet, his identity protected by the fact that until he died he was always only referred to in scientific publications by his initials HM. What researchers found was that Henry's working memory was largely intact. If he had been given the task I gave you at the start of this piece he would have performed just like you. In addition, Henry's long term memory, up until the point of his operation was also largely intact as well – although not perfectly, as he forgot some fairly major events from his pre-operation life. By and large, however, his pre-surgery long term memory was relatively unaffected. What seemed to be deficient was the ability to create new long-term memories. He had lost the ability to transfer information from working memory into long term memory. From the point of his surgery onwards Henry was stuck in a permanent present that lasted only 30 seconds!

In the 1990s Henry was eventually scanned to locate the exact position of the parts of the brain he had lost. It turned out that the surgery had bilaterally removed a major part of the temporal lobe called the hippocampus. It was due to studies of Henry, and other patients who had undergone similar surgeries, or endured brain damage, due to traumatic accidents or brain infections, that the hippocampus was located as the primary site in the brain where items in working memory are consolidated into long-term memory. Note that it isn't the place where long-term memories are stored. If it was, then Henry would not only have lost the ability to create new long-term memories but he would have lost all his old ones as well. In fact where and how our long term memories are located is still not very clear, but the cortex of the brain appears to be important in this respect. What is becoming increasingly obvious, however, is that the way most of us think about long-term memory – as a kind of library in the brain, or like the memory on a computer with folders and files, is probably not how human long-term memories are organised. One thing is certain, however: our long-term memories do not reside in the hippocampus. Moreover, the hippocampus doesn't seem to be necessary to retrieve the memories that have been previously stored, back into consciousness. But what it does appear to be absolutely essential for is that initial processing step which consolidates an item in working memory into an item in long-term memory. It has been described as the 'hub' that binds together information into memories.

We have therefore located the site in the brain where this critical process occurs. The key question now is how does it happen, and is there anything we can do to affect how solid the transfer process has been. Can we make the transfers more permanent when we need to? The answer is yes. It has been known for some time that electrically stimulating the hippocampus while learning may improve memory functions. The main problem is that the hippocampus is a structure that is deep inside the brain, and so to stimulate it electrically it is necessary to perform some surgery to implant small electrodes into the appropriate region. Clearly this is not something that could be done on a routine basis. However, there are other ways to stimulate the brain, one of which is called transcrainial magnetic stimulation or TMS. This uses magnetic fields to stimulate specific brain areas and involves placing a large plastic pad on the head adjacent to the area that needs to be stimulated. The technique has been used since the 1990s and it seems to be effective as a cure for some types of depression. Unfortunately, the magnetic stimulation cannot penetrate deep enough into the brain to directly stimulate the hippocampus. When long term memories are formed it is thought that the hippocampus interacts with areas of the cortex which are closer to the brain surface and hence accessible to TMS. A group of researchers at Northwestern university in the USA, led by Joel Voss, have now tested to see if stimulating a region of the cortex, known as the parietal lobe, during a memory task enhances the consolidation of the long term memory. Their paper was published in the journal *Science* at the end of August.

The way they located the exact area to stimulate was to perform a functional MRI scan which located the exact region of the parietal lobe cortex that lit up and was connected to the hippocampus during a memory task. They studied 16 healthy adults aged between 21 and 40 and they controlled the experiment in three different ways. The first control was they gave half of the subjects sham stimulations. It is actually impossible to detect the magnetic field so subjects didn't know if they were getting the stimulation or not. They were stimulated (or sham stimulated) for 20 minutes a day for 5 days. They then used functional MRI to see whether the stimulation had enhanced the link between the cortical area and the hippocampus. Because they only stimulated the left side of the brain the second control was to look at both right and left sides in the fMRI images. They found that the stimulation significantly enhanced connectivity between the cortex and the hippocampus only on the left side, and only in the stimulated subjects.

Great: more connectivity. But did it positively affect their long-term memory? The researchers used a test called 'face cued word recall' to test the subjects memory. This involves showing people pictures of peoples faces arbitrarily linked with a printed word. Later the subjects are shown only the face and they have to remember the associated word. At the end of treatment the stimulated group had a significantly enhanced memory compared to the sham stimulated individuals. I have seen it reported elsewhere that the improvement was 30%, but from the actual paper I can't see how this number was calculated. Nevertheless it is clear that the performance was better in the stimulated individuals. Perhaps most interestingly the extent of the change in memory performance was positively associated with the extent of change in connectivity between the stimulated area and the hippocampus. Finally, to show that the targeted nature of the stimulation was important they repeated the work but this time stimulating the motor cortex which does not have strong links to the hippocampus. This third 'control' stimulation had no effect on memory.

How long do these stimulated memories last? At present we don't know. TMS however is considerably easier to use than deep brain electrical stimulation because all you need to do is place a plastic pad on the side of your head. It appears to be safe, and the effects are very clear: you can enhance your long term memory using external magnetic stimulation. If the memories really are consolidated better, it may not be long before you can buy designer headgear with TMS units in them that you plug in and wear while revising for your exams to improve your performance.

Cheetahs show why walking is more expensive than running

October 2014

Usain Bolt is the fastest human being ever to run along a sprint track. His dramatic appearance on the world stage in the 100m race at the Beijing 2008 Olympic games, having never previously won a 100m sprint in international competition, is one of the greatest sprint race wins of all time. In the closing stage of the race, Bolt was meters ahead of the rest of the field and actually slowed down as he crossed the finish line. More amazingly, it later turned out one of his shoe laces was untied! Yet he still finished in 9.69 seconds and broke the world record. Shortly after the games, analyses of his Beijing run by a team of astrophysicists from the University of Oslo in Norway suggested if he had not slowed down he could have run the race in under 9.6s. A year later there was no need to speculate on what could have been. Bolt won the final of the 100m sprint in the world championships held in Berlin in 9.58s and smashed his earlier world record. Yet, this is still not the fastest he has run 100m. When sprinters engage in a 100m race they have to start from standing still. So the first 20m involves accelerating to their top speed. In a 150m race, however, when the person hits the 50m mark, they are already running as fast as they can. In a 150m race held in Manchester, England in May 2009 Bolt ran the last 100m in a staggering 8.7 seconds. This remains the fastest that any human has run 100m, and is equivalent to a speed of 41.4 km/hour.

So how would Usain Bolts performance compare to a cheetah? Cheetahs are commonly suggested to be the fastest land mammals. They are often described as the Ferrari's of the animal world. Estimates of how fast cheetahs actually run, however, are difficult to make in the field, and estimates of their top speed have varied from 112-120 km/hour. Moreover, the field situation is somewhat more obstacle ridden than the average track and field course. Imagine how fast Usian bolt would have run the 100m if half way along the track he had to swerve around a bush and avoid a large boulder. In 2012, Cincinnati zoo, in the USA, decided to test exactly how fast a cheetah could run under the same conditions faced by Usain Bolt. On June 20th they took five cheetahs to a running track and recorded how fast they could run the 100m sprint. To get the cheetahs to sprint they pulled a fluffy toy dog along the track on a high-speed cord. Some of the cheetahs were less than engaged in the exercise and put in times around 10s. But one cheetah, called Sarah, put in a blistering run of 5.95s, averaging a speed of just under 60.5 km/hour. To put that in context, when Bolt ran his 2008 world record in the birds nest stadium, after 5.95s he had just passed the 55m mark. Sarah the cheetah would have beat him by 45m! And remember this is a well fed zoo cheetah mucking about chasing a fluffy toy dog. Catherine Hiker of Cincinnati zoo who organised the cheetah 100m sprint event speculated that in the wild where a hungry cheetah would have to chase down its dinner, the attained speeds might be much faster, possibly consistent with the previous speed estimates of up to 120 km/hour.

The real problem with these speeds reported from the wild however is that it is difficult in the field to get good estimates of times and distances that the animals travelled – so the reliability of the estimates is questionable. That problem was solved earlier this year by team from the Royal Veterinary College in London, UK who fitted five wild cheetahs living in Botswana with inertial measurement units and accurate GPS units that enabled the precise hunting paths and speeds of the animals to be tracked during 367 chases. Their work was reported in the journal *Nature* (Wilson et al 2014 Nature 498: 185-189). The detailed examination of the chases revealed that the average distance that cheetahs ran during chases was 173m and their longest chases varied between the individuals from 400 to 560m. On average however the speeds attained by the animals were slower than the zoo animals at the track. The mean top speed was only 14.9 m/second compared to an average of 16.8 m/second for the cheetah Sarah, over her 100m sprint. But this average speed in the wild hides a wide range, and the fastest speeds in the field were much higher. The fastest recorded speed was actually 25.9 m/second (93 km/h) equivalent to covering 100m in 3.86 seconds! At that pace the cheetah would be crossing the finish line before Usain Bolt got 40m down the track. These cheetahs were mostly chasing impala. The authors speculated that Cheetahs in east Africa which specialise on Thompson's gazelle may perform even faster.

Running at such speeds requires an enormous power input. Using kinetic energy calculations from the inertial monitoring system it was estimated that the cheetahs muscles must have been delivering around 100 watts/kg body weight when sprinting at speeds of 10-18 m/s. In fact a biomechanical analysis of Usain Bolts record winning run in Berlin in 2009 indicates he only managed to achieve a quarter of that level of power output. This raises the question of how cheetah achieve these phenomenal power outputs. Perhaps there is something special about their muscles. The same team from London, has just published another paper (West et al 2014 Journal of Experimental biology 216: 2974-2982) addressing this exact problem. They took muscle samples from a dead cheetah, shortly after it had been euthanized, and looked at the ability of fibers extracted from the muscle to generate power during electrically stimulated contractions. For comparison they used muscles taken from rabbits. The power generated from contraction of the cheetah type 2 fiber segments was 92.5 watts/kg, but remarkably the corresponding output from the rabbit muscle was 119 watts/kg. The superior locomotor performance of the cheetah therefore remains an enigma, but it is clearly not due to greater contractile power generation by their muscles.

A presumed consequence of these high energy demands has been that cheetah must live on a continual knife edge of energy supply and demand. They might therefore be particularly susceptible to prey stealing by larger predators such as lions and hyenas. In fact cheetah populations have declined from around 100,000 at the turn of last century to just under 10,000 today. The last Cheetahs were observed in Arabia in the 1950s, and the last 3 known in India were shot by the Maharajah Ramanui Pratap Singh Deo in Central India in 1952, and now it is almost completely confined to Africa, apart from a small population in Iran probably numbering less than 100. It has been widely speculated that the decline in populations throughout Africa has been caused by tourism activities that encourage large lion and hyena populations. As such lion and hyena populations increase, the likelihood that cheetahs will get their hard earned prey stolen also increases, and because they live on the edge of an energetic precipice, it is pretty easy for such food stealing to push them over the edge. In fact a paper in Nature over 15 years ago (Gorman et al 1998 Nature) showed that this was very probably a key problem for the African Wild dog. Gorman's team used a method called the doubly-labelled water technique to measure the energy costs of free-living wild dogs. This technique involves injecting the animals with stable isotopes of oxygen and hydrogen, and then tracing their elimination.

The differential elimination of the two isotopes is related to energy expenditure. These direct measures showed that the energy demands of free-living wild dogs are incredibly high. As a multiple of resting metabolic rate the levels were around 5.2x the resting rate. This compares for example to humans that expend energy at about 1.6 to 1.7x resting. By incorporating these direct measurements of the energy expenditure into a mathematical model it was shown that if the rate of food stealing by lions and hyenas for these dogs was to increase to 27%, then to cover their losses the dogs would need to increase their hunting effort to cover the entire day, increasing their energy expenditures to completely unsustainable levels.

Because cheetahs are about the same size as African wild dogs and are also occasionally observed to have their prey stolen by lions and hyenas it has always been thought that they must also be susceptible to the same sorts of problems, and that this has been a major contributory factor to their declining population. Now a new study published in *Science* by Michael Scantlebury and colleagues from Queens University in Belfast, UK, has used the same method that was used on the wild dogs to measure their energy expenditure, combined with a similar GPS technology used by Wilson and colleagues to study the hunting dynamics of cheetahs, and the results are somewhat surprising. The first surprise was that actually the daily energy demands of cheetahs are quite unremarkable. As a multiple of resting metabolic rate they were only expending energy at about 1.6x their resting rates. This is about the same as that observed in humans, and substantially less than in the wild dogs. Cheetahs may be the Ferraris of the world, but like most actual Ferraris it seems they spend their time mostly driving around slowly in traffic. Plus although these cheetahs engaged in different numbers of chases each day, this did not explain any of the day to day differences in their energy demands. The only factor that came out as important in driving their daily energy demands was not the distance (or time) they spent chasing prey, but the distance (or time) they spent walking. This was a far more significant factor driving their daily energy requirements.

Walking it seems is far more expensive for cheetahs than running, basically because the time they spend walking is much greater than the accumulated time spent in the dramatic short chases after prey items. As Wilson and colleagues noted in their study in Botswana, cheetahs on average only engaged in 1.3 hunts per day covering about 250m, while they walked on average 6km each day. Walking 6km costs considerably more than sprinting 250m. Similarly, Scantlebury and colleagues observed their cheetahs engaged in an average of 1.2 hunts per day, lasting a total of only 38s, but they spent 2.86 hours walking. In fact this pattern is probably not something unique to cheetahs. Usain Bolt probably spends much more energy and travels a greater distance over the course of a competition day just walking about, than he does during the 9.6s spent sprinting 100m down the track. It is just we have never really thought about cheetahs in this way before – instead always focussing on the dramatic high cost chases, and forgetting how short they actually are compared to the far less spectacular, but quantitatively more important, time they spend wandering around.

These findings have some important consequences. The first is that because the cost of chasing down prey is actually quite modest, the implications of losing prey to food stealing by lions and hyenas is much less serious than it is for the wild dog. In fact when Scantlebury and colleagues put their data into the identical mathematical model used previously to explore the impacts of food stealing on wild dogs they found that the same increase in food stealing, to 27% of items, that would be devastating to the wild dogs, had only a minor impact on the cheetah energy budgets – increasing their time spent walking from 2.86 to 3.9 hours per day. In fact cheetahs would have to lose substantially more than 50% of their prey to food stealing before the effects become close to being serious. Since such levels of food theft are never observed in the wild the implication is that the declines in cheetah populations probably have nothing to do with competition from larger predators such as lions and hyenas. Rather the more important factors that influence cheetah energy budgets, are things that affect the distances they need to walk each day – and chief among these are man made barriers in their environment such as fences. In turn it seems probable that these human influences are much more important factors causing the declining cheetah populations.

Do you need to be smart to make art?

November 2014

One of the more remarkable things to have been revealed during the last year was the revelation in April that ex-US President George Bush (junior) paints portraits in his spare time. He recently exhibited pictures of 24 world leaders, whom he had mostly met in person during his time in office, in an exhibition at the Bush presidential library in Dallas, Texas, USA entitled 'The art of leadership'. He said that he had done the portraits in the spirit of friendship. Why is this a surprise, or even a shock, as it seems to be? The reason it is a surprise is that many people regard 'art' as an intellectual pursuit. You need only read the descriptions of many contemporary art exhibits to realise that some artists themselves certainly see what they are doing as an intellectually charged activity. For example, I recently attended an exhibition in the Faurschou gallery in the 798 Art district in Beijing, which is one of the cities major contemporary art exhibition spaces, and what was on show was an exhibition called 'We the people' which basically consisted of some large structures made of copper. There were about ten of them spread around the room and I wandered around them for a while trying to see whether they had been arranged in some special way. When I read the text next to it, it turned out that they were actually life sized replicas of pieces of the statue of liberty in the USA. Not only that, which was surprising enough, but apparently something I had missed was that the work raised questions about the subject of freedom, including whether or not it is necessarily everyone's human right, as well as its fragmented nature reflected the artists own fragmented life and what it means to be an immigrant. Really? Wandering around them I hadn't quite picked that up. Clearly, however, the artist Danh Vo, had felt that he was making some major intellectual statement as he constructed them and then spread them out. And he is not alone. Art critics have rushed to positively review this work as it has travelled the globe. In New York, for example, where the original statue is based, art critic Karen Rosenberg of the New York Times said "Once you get past the glaring obviousness of its central metaphors, Danh Vo's, 'We the people' is compelling".

So the discovery that George W Bush is also an artist poses a major dilemma for the art world, because when he was in office George W was universally derided as being a moron. Even George Bush himself recognised that he was not among the intellectual elites of this world. He famously once addressed a graduation class at a major US university by saying "Congratulations to the first and second class students here today. And to the third class students, well you too like me may one day become President of the United States." So how is it possible that that someone, so clearly not an intellectual, can produce 'art'? It poses an interesting question. Or as a description of a contemporary art work might put it. His work fundamentally questions our underlying assumptions relating to the production of art, its central metaphorical themes and whether it is related to intellectual capacity. There are only 3 solutions to this dilemma: George is in fact not as stupid as he was thought to be, or what he is doing isn't really art, or he is doing art and he is a moron, but you don't have to be smart to produce art.

The art world has universally gone for the second solution. Jason Farago, art critic at the Guardian newspaper in the UK described the portraits as 'vacant, stubborn and servile. They are art that tells us nothing at all'. Unlike, for example, a collection of life sized bits of the statue of liberty, I imagine, that tell us all about what it is like to be an immigrant, once you get past the glaringly obvious central metaphors. They are, he further opined, "paintings by an artist anxious, or perhaps incapable, of doing anything that might leave a mark." His views are mirrored by almost the entire art world. I could probably write four pages of text simply repeating obnoxious remarks. Apart from simply deriding the paintings as rubbish, a further approach has been to compare the paintings George Bush has produced to the 'paintings' made by animals. This reminds me that I once bought a painting done by a dolphin. It was in the 1990s and we were doing some research at the Dolphin Research Center in the Florida Keys in the USA. The dolphins there routinely did 'paintings'. In fact one was on sale at the center for about 1000 US dollars, called something like "Apollo's last message". The Dolphin Research Center in the Keys is a pretty remarkable place because the dolphins are kept in 'pens' that are actually just cordoned off bits of the sea. The fences that hold them in are only about 6 inches above the surface, so the dolphins are completely free to come and go as they wish – they stay presumably because of all the free fish. When a hurricane is coming they take down all the barriers completely and encourage the dolphins to swim away somewhere safe. The 'Apollo's last message' was a painting done by a dolphin which had subsequently gone missing in a hurricane, and had not come back with all the others, so they presumed it had died, and wouldn't be doing any more art for the foreseeable future. The picture had a smudge of black paint at the bottom of it 'Was this Apollo trying to tell us about some dark things about to happen in his future?' the bit of paper that also said '1000 dollars' on it asked us to speculate. Well, clearly more likely was that someone had placed a paint bush covered in black paint in Apollo's beak and then tried to make him swim past a piece of paper. Apollo probably had no idea he was even painting, let alone foretelling his bleak future. I couldn't afford Apollo's last message so instead I bought a T-shirt painted by another dolphin. I only wore it once. Back home in the UK my wife said it kind of looked like I had been painting the kitchen and dropped the brush down myself. Comparing what George Bush has produced to these animal creations is a phenomenal insult. If I look at what George has produced then it is clear who each of the paintings represents. If I look at the T-shirt my dolphin painted I am hard pressed to know if someone didn't just drop a brush on it accidentally. But such is the challenge that 'George Bush the artist' represents, that this is the sort of level to which the argument that what he is doing cannot be classed as art has sunk. Paula Young Lee from Tufts university in the USA, writing for the online magazine 'The Conversation' likened him to an elephant she had observed painting in Thailand. The elephant the author claimed isn't doing 'art' it is, like George, just 'painting stuff'.

A more recent critique of the paintings is that he copied them from pictures on the internet downloaded using google image, or from their Wikipedia entries, and hence they are fakes. Two art critics published this exposé shortly after the exhibition was first opened, as if to say – so that's it, they are paintings done from downloaded pictures and that isn't 'art'. But who says you have to paint a portrait from a live sitting, or from memory, for it to be 'counted'. American artist Leon Golub painted around a hundred portraits of world leaders in the late 1970s and it seems likely these were also done from photographs since he didn't actually meet any of them. Yet nobody has thought to condemn these as fakes. Is it really any harder to paint if the person sits there in front of you, than it is if you have their photograph. Plus some of the claimed 'copies' seem rather far-fetched anyway. I mean if they look anything like the person they are depicting, then they are going to look like a photo of the person as well. In contrast to this deluge of negativity, a teacher who taught him how to paint after he left office has said that he is passionate about his painting, and has a Rembrandt hidden inside him!

So none of this resolves the key question that George Bush's paintings open up – do you need to be smart to make art? Another way to look at this is to ask when in our evolutionary history did we start doing it, compared, for example, to when we started doing other activities that require a degree of sophisticated intellect like growing crops, or writing things down, or doing maths. Since on average our brain size and our presumed intellect has expanded over time, then the earlier something appears in our history, then the less intellect on average it might require to do it. Until recently the earliest examples of 'art' were cave paintings from northern Spain and France. These depict animals being hunted and have been dated to between 16,000 to 40,000 years ago. The conventional wisdom has been that art originated in Europe and then slowly spread to other places. Other paintings have been found elsewhere, notably in Indonesia and Australia where many paintings, and in particular hand stencilled drawings were first discovered in the 1950s. These paintings were universally regarded as much younger in origin. However, a recent paper published in the journal *Nature* has turned this view of the origins of art on its head. The study involved an analysis of some hand stencilled paintings from the island of Sulawesi in Indonesia. The investigators based at Griffith University, in Australia, noticed that on top of the paintings some rock nodules had developed called 'cave popcorn'. Since these nodules had to have appeared after the paintings were made, the authors argued that by dating the 'popcorn' they could estimate the most recent time that the paintings had been done. To estimate the age of the nodules they used an isotope technique which looks at the uranium content as well as the uranium decay products. Put simply, the longer something has existed the lower the uranium level and the greater the levels of its decay products. And the answer from these uranium based dates was a surprise! It turns out these paintings are at least 40,000 years old – contemporary with the European cave paintings made 8000 miles away. This suggests that either artistic ability developed independently in the two populations around the same time, and hence there were at least two cradles of artistic discovery. Or, as the lead author Maxime Aubert favours, the origin of artistic endeavour was common to them both, and was probably much earlier around 65,000 years ago when modern humans first started to migrate out of Africa. We just haven't found any ancient art in Africa yet, because its geology is far less congenial to preserving it. Almost simultaneous to the study in *Nature*, another report published in the *Proceedings of the National Academy of Sciences* of the USA has suggested that some scrapings on the walls of a cave in Gibraltar on the Iberian peninsula in Europe may have been 'art' made by Neanderthals.

Whatever way you look at it, however, 'art' is a very early aspect of human evolution. We were probably making art 40,000 years before we started to cultivate plants and invented agriculture, and 60,000 years before we started to write and represent things symbolically, which led to the invention of mathematics in around 2000 BC. Making art is a primitive skill, and on that basis you probably don't need to be particularly smart to do it. George Bush being a painter is not an anomaly that needs to be somehow explained away by comparing him to a performing elephant. He is just someone who brings the reality starkly to light, because of all the years he spent being mocked as an idiot. George himself has been stoical in the face of the criticism of his paintings. He is quoted as saying, when someone pointed out the surprise people have that he could paint, "Of course, some people are surprised I can even read". Perhaps there is a comedian hidden in there as well. Oh no, don't tell me you have to be an intellectual to be one of those too!

Pandas – dead end or dead wrong?

December 2014

Shortly after I arrived to live in China, in 2011, there was an amusing item on the TV evening news. A giant panda had wandered into a small village in Sichuan province and when cornered by a group of villagers it had bitten one of them, before running back off into the countryside. Of course, being bitten by a wild animal is not so amusing, but what I thought was very funny was the bitten mans reaction to it. He was interviewed by the news team from CCTV, and he stood there with an enormous bandage on his hand, holding it up so the camera could see it, and he said "It is my honor to have been bitten by the Chinese national symbol." I don't think you can get much better proof of the fondness that is felt nationally in China for the panda. The panda has effectively replaced the dragon as the national animal, and a panda based cartoon was one of the five mascots of the 2008 olympics. This affection is felt not just in China but across the world. Shortly after the biting incident, China loaned two pandas to Edinburgh zoo, close to where I live in Scotland. I had chance this summer to meet and have lunch with Ian Valentine who is the curator at Edinburgh zoo responsible for the pandas. He told me that the zoo had seen a significant increase in its income after the pandas had arrived. In fact, he said, for zoos worldwide having a Panda is the second most attractive exhibit that they can have, in terms of generating interest in the general public, and increased income for the zoo itself. This is even accounting for the 1,000,000 US dollars per year that zoos have to pay the Chinese government for a panda loan. In terms of visitor numbers and finance, having a panda in a zoo is exceeded by only one other thing. And that is having a baby panda! In Scotland over the last summer the news was dominated by just two items. The first was whether Scotland would leave the UK to become independent. A close second was whether the female panda in Edinburgh zoo 'Tian tian' was pregnant or not. In the end the two stories had similar outcomes. Scotland looked like they would leave but didn't. Tian tian looked like she was pregnant, but wasn't.

Given such international interest in the giant panda it would seem an absolute given that we should attempt to preserve pandas in the wild to prevent them from becoming extinct. The sub-text of the worldwide zoo program is that pandas are being bred, not because they draw in literally millions of dollars of income for the zoos, but that ultimately the breeding populations in zoos may provide a reservoir of animals that can be used to restock the wild population. At present there are over 350 in captivity, with maximally about 3000 in the wild. Conserving the panda has been a Chinese, and international, priority for at least the last 40 years, since the populations in the wild were reduced by over 1000 individuals between the 1950s and 1970s. Indeed the world wide fund for nature (originally the world wildlife fund: WWF) has as its symbol the giant panda. It may come as some surprise therefore to find out that there is a significant body of opinion among conservationists that trying to save the giant panda is a waste of time – and more importantly a dangerous waste of money, which could be used for more important conservation goals.

The suggestion that saving the panda is a waste of time and effort dates back about 5 years or so. The argument is that pandas are basically in the state they are in because they have gone into an evolutionary cul-de-sac from which they cannot escape. Put bluntly, extinction of the giant panda is inevitable, but it is basically their own fault. They have set themselves on a collision course with oblivion. The first test of whether a species is worthy of a conservation effort is that they must have some sort of self preservation instinct. Pandas, it is argued, abjectly fail this test on two fronts. First, their breeding habits make it almost impossible for them to procreate effectively. The female comes into oestrus once a year for only a few days, making the likelihood that a male will be around to inseminate her minimal. This is combined with the fact that male pandas seem to have a pathological disinterest in sex. The difficulty of getting pandas to breed has reached almost mythical status. Zoos in the past have resorted to giving male pandas Viagra, and showing them videos of other pandas mating – although it is unclear if the intention here was education or arousal. Even once a female panda gets pregnant their offspring are born at an incredibly small size for the body weight of the mother. For example, a human mother weighing about 70 kg typically will give birth to a child weighing between 3 and 4 kgs. A 100kg giant panda however gives birth to a baby weighing only about 150g. Although marsupials give birth to proportionally smaller offspring, which then live in their pouch, the panda gives birth to the smallest offspring proportional to its body weight among the Eutherian mammals. This small size makes them highly vulnerable and very often if the mother has two cubs she abandons one of them to die.

The second problem is their dietary habits. Pandas are carnivores who have become vegetarians. Although they have retained large canine teeth, and a short alimentary tract, that are ideally suited to killing things and digesting them, pandas eat almost nothing but bamboo. And bamboo is a plant that has really low nutrient value, that the panda finds almost impossible to extract, because it has the guts of a lion. The result is that a panda needs to spend most of its waking hours feeding. Not only that, but bamboo species have a semelparous reproductive strategy, which involves a periodic flowering followed by widespread dieback of the plants. So in essence pandas have evolved to become specialists on a plant that they are very poorly adapted to digest, and which every 40-100 years becomes completely unavailable – leading to documented increases in their mortality, from which they cannot recover because of their low reproductive rates. This problem of repeated population bottlenecks, following bamboo die offs, is suggested to have reduced their genetic diversity, and hence their capacity to respond to environmental change. This has been suggested to underpin their long term population decline. They are basically becoming progressively more and more inbred. These arguments led British author and wildlife activist Chris Packham, to conclude in 2009 that conserving "the panda is possibly one of the grossest wastes of conservation money in the last half-century." More extremely he said that he would "*Eat the last panda!!* if I could have all the money we have spent on panda conservation put back on the table for me to do more sensible things with". (Although he later apologised for this latter comment). When the Washington zoo pandas gave birth to a baby panda cub in 2013 Timothy Lavin editorial board member at Bloomberg View wrote an article summarising many of these arguments entitled 'Why I hate pandas and you should too'. His piece started "Congratulations Washington zoo on your new panda cub. You're prolonging the existence of a hopeless and wasteful species the world should've given up on long ago."

More recently, in early 2014, a survey of 583 biological scientists conducted by Dr Murray Rudd from the University of York in the UK found that a majority

(60%) of them considered that it would be a good idea to focus limited conservation effort and resources on species that can realistically be saved from extinction and simply give up on the rest as a waste of time: a so called conservation triage. Among the species in the 'too expensive to save' category was the giant panda. Wildlife biologist Paul Goldstein has pointed out that choosing how to allocate resources for conservation is currently largely done on the basis of how much affection people have for them. Top of that list comes the panda, and perhaps also the tiger. It is however he argues a poor reason for choosing how to allocate conservation resources. Baby pandas look incredibly cute, but does that make them more deserving of our resources for conservation than say sharks, populations of which have been decimated. Emotional links to the animals he argues obscure our ability to make more rational decisions. The point is, does it make sense to invest so much time, effort and money on conserving the panda when it is an animal doomed to extinction no matter what we do. The argument is not only whether it is sensible but whether it is fair on all the other species deserving our protection which are more worthy cases because they could feasibly be saved.

However, conservation biologists and in particular panda biologists have started to resist these arguments. With respect to the giant panda these counter arguments have been beautifully laid out in a recent article led by Fuwen Wei, deputy director of the Chinese

Academy of Sciences Institute of Zoology in Beijing and a veteran panda biologist, who discovered that the wild population is probably much larger than we thought – based on collection of feces in the wild and identification of unique individuals from extracted DNA. The article was published in Molecular Biology and Evolution and involves clear rebuttals of the major arguments that pandas are a dead end doomed to extinction and hence unworthy of conservation efforts.

With respect to the bamboo diet, Wei and colleagues point out that the fossil record suggests pandas adopted this diet between 2 and 4 million years ago. Although it is often stated they are poorly adapted to this diet, in many respects they are very well adapted. They have enlarged zygomatic arches on their skulls which accommodate enlarged zygomatic-mandibular muscles that are specialised for chewing the hard bamboo stems, and which incidentally gives their face the large rounded appearance. Their teeth are not just carnivoran, as is often stated, but actually are large and flat with elaborate crown patterns for effective crushing of their diet. Critically, they have an extra pseudo-thumb which is an enlarged radial sesamoid bone which allows them to effectively grip and manipulate bamboo shoots Moreover, the panda has a large number of genetic mutations in its genome that physiologically adapt it to this specialised diet – including for example the loss of a taste receptor for umami that is important for sensing the taste of meat. Not only this, but their specialised gut microflora allows them to more efficiently digest the bamboo, despite their genome lacking enzymes to digest cellulose. During some bamboo diebacks it was observed that many pandas died, but this was basically because such panda populations had been forced into areas with few alternative sources of food. In other areas facing bamboo dieback pandas simply switch their diets to other non-flowering bamboo species. This is probably what pandas have done since they first specialised on bamboo and has enabled them to survive for more than 2 million years on this food source involving more than 20,000 bamboo diebacks. It is only when human activity forces them into high mountain areas with limited numbers of bamboo species that they start to struggle. Wei et al also summarised recent genetic work showing that the panda population is not in long term decline as is often suggested to result from the supposed bottlenecks after bamboo diebacks. Consistent with the suggestion that the population probably did not historically die along with the bamboo the reconstructed history shows contractions of the population coincide with glaciations and global cooling events between which the populations expand. There is no evidence of low genetic diversity and indeed most population studies indicate that the conservation effort is working, because wild panda populations are increasing.

With respect to reproductive behaviour the paper shows that the key problem with panda reproduction is not the panda. In the wild recent work suggests females give birth every 2 years or so. Recent estimates indicate a reproductive rate of around 63% per annum in two wild populations and that there is very high cub survival. This completely contrasts the impressions gained form attempts by zoos to breed them. Clearly the problem was the zoos and not the pandas. As our knowledge of panda breeding biology has increased captive breeding success rates have also started to climb. Indeed cub survival in modern breeding facilities is over 90%. The suggestion that the panda is an evolutionary dead end is simply dead wrong. The panda is well adapted to its diet, has good reproductive potential and high genetic diversity. The key issue the panda faces is massive human development and human interference. Wei Fuwen's paper ends with a sentence bordering on poetry: "The giant panda has not persisted despite its evolution; it has managed to hang on despite ours."

Finally, as Diane Walkington, the WWF Head of species, pointed out earlier this year, in the end we only conserve species by conserving the habitats they live in and so attempting to prevent the giant panda from going extinct is actually an attempt to preserve the bamboo forests where they live. This results in preservation of a whole ecosystem of animals, such as dwarf blue sheep, ground pheasants and the crested ibis, for which the panda is just a figure-head. And figure-heads are useful because they act as a focus for fund raising. A drive to raise money save the bamboo forest, however well intentioned, is unlikely to raise as much funds as a drive to save the panda. The point is that the pot of money for conservation is not fixed. If there was actually a mechanism to decide that no further effort would be devoted to panda conservation, then the money that people donate for panda conservation could not be reallocated to some other cause, simply because that money would likely no longer be donated to conservation efforts at all. Attempting to conserve the panda allows us to save lots of other animals that most people have never heard of, and probably care even less about, but they form an essential component of the ecosystem in which the panda lives, and that we need to preserve, to save them all.

Original sources and further reading

January   Tool use – how low can it go?
Dinets, V. et al (2013). Crocodilians use tools for hunting. *Ethology, Ecology and Evolution* 27: 74-78.

Dinets, V. (2013) *Dragon songs: love and adventure among Crocodiles, alligators and other dinosaurs relations.* Arcade Publishing

February   Fears from the past
Dias, B.G. and Ressier, K.J. (2014) Parental olfactory experience influences behavior and neural structure in subsequent generations. *Nature neuroscience* **17**: 89-96.

March   Cheap formations

Portugal, S.J. et al. (2014) Upwash exploitation and downwash avoidance by flap phasing in ibis formation flight. *Nature* **505**: 399-402.

April   Flashing too early

Neuman, S.B. et al (2014) Can babies learn to read? A randomised trial of baby media. *Journal of Educational Psychology* **106**: 815-830.

May   Lazy? It could be genetic, and there could be a cure around the corner!

Zhang, Z., *et al.* (2014) Mutation of SLC35D3 Causes Metabolic Syndrome by Impairing Dopamine Signaling in Striatal D1 Neurons
PLOS Genetics **10**: e1004124

June **Crossing the boundary**

Hodges, A. (2012) *Alan Turing: The Enigma.* Vintage digital. 768 pp.

Mason F. (2014) *Turing: the tragic life of Alan Turing.* Bookcaps study guide. 76 pp

Cawthorne, N, (2014) *Alan Turing: The enigma man.* Arcturus publishing. 131 pp.

July **Youthful blood**

Villeda, S.A. *et al.* (2014) Young blood reverses age-related impairments in cognitive function and synaptic plasticity in mice. *Nature medicine* **20**: 659-663.

Katsimpardi, L. *et al.* (2014) vascular and neurogenic rejuvenation of the aging mouse brain by young systemic factors. *Science* **344**: 630-634.

Sinha, M. *et al.* (2014) Restoring systemic GDF11 levels reverses age-related dysfunction in mouse skeletal muscle. *Science* 344: 649-652.

August   Tibetans have ancient genes for running in thin air

Huerta-Sanchez, E. *et al.* (2014) Altitude adaptation in Tibetans caused by introgression of Denisovan-like DNA. *Nature* **512**: 194-197.

Beall, C. *et al.* (2010) Natural selection on EPAS1 (HIF2 alpha) associated with low haemoglobin concentration in Tibetan highlanders. *Proceedings of the National Academy of Science* **107**: 11459-11464.

September   Stimulating memories

Wang, J.X. et al. (2014) targeted enhancement of cortical-hippocampal brain networks and associative memory. *Science* **345**: 1054-1057.

Corkin, S. (2013) *Permanent present tense: The man with no memory and what he taught the world.* Penguin books, London, UK.

October   Cheetahs show why walking is more expensive than running

Scantlebury, D.M., *et al.* (2014) Flexible energetics of Cheetah hunting strategies provide resistance against kleptoparasitism. *Science* **346**: 79-81.

Wilson J.W. et al (2013) Cheetahs, *Acinonyx jubatus*, balance turn capacity with pace when chasing prey.. *Biology Letters* **9**: e20130620

Wilson, A.M. et al (2013) Locomotion dynamics of hunting in wild cheetahs. *Nature* **498** 185-190

West, T.G. et al (2013) power output of skinned skeletal muscle fibres from the Cheetah (*Acinonyx jubatus*). *Journal of Experimental biology* **216:** 2974-2982.

November   Do you need to be smart to produce art

Aubert, M. et al (2014) Pleistocene cave art from Sulawesi, Indonesia. *Nature* **514**: 223-226.

December   Pandas – dead end or dead wrong?

Wei, FW et al. (2014 : in press) Giant pandas are not an evolutionary cul-de-sac: Evidence from multidisciplinary research   *Molecular   biology   and   evolution* doi:10.1093/molbev/msu278

*John Speakman was a co-author on the papers by Scantlebury et al (2014) and by Zhang et al (2014)

www.ingramcontent.com/pod-product-compliance
Lightning Source LLC
Chambersburg PA
CBHW051726170526
45167CB00002B/812